Edward Bovy

Edward Bovy

This book is dedicated to the memory of June Mowatt, a wife, mother and noted geologist who spent much time along the Dalton Highway. June's serious illness prevented her active participation in the preparation of the text; however, her moral support motivated all who worked on the book.

Riches from the Earth

A Geologic Tour Along the Dalton Highway, Alaska

Bureau of Land Management and
The Alaska Natural History Association
Anchorage, Alaska

Acknowledgements

Field geologists and writers
William R. Diel, Branch of Mineral Assessment
Arthur Banet, Jr., Branch of Mineral Assessment

Editors
Robert Bascle, Branch of Mineral Assessment
Gary Brougham, Branch of Mineral Assessment
Tom Mowatt, Branch of Lease Operations

Project Concept
Gary Brougham, Branch of Mineral Assessment

Graphics
David Evans, Branch of Mineral Assessment

Design
Edward Bovy, Office of Public Affairs

Photos
Front cover: Aufeis near outlet of Galbraith Lake (Edward Bovy)
Preceeding pages: Mt. Dillon near Mile 207 (Edward Bovy);
Granite tors at Finger Mountain (Edward Bovy)
Back cover: Brooks Range near Toolik Lake (Edward Bovy); young grizzly bear along road in Brooks Range (Arthur Banet, Jr.).

Alaska Natural History Association
605 West Fourth Avenue, #78
Anchorage, Alaska 99501

This publication is available by mail for $9.95 each plus $3.00 shipping and handling per order from the above address.

The Alaska Natural History Association is a non-profit organization dedicated to enhancing the public's understanding and conservation of Alaska's natural, cultural and historical resources by working in cooperation with land management agencies and other educational organizations throughout Alaska.

The BLM-Alaska Branch of Mineral Assessment takes this opportunity to thank the following people whose support made this publication possible: Hillary Oden, Assistant Director for Energy and Minerals Resources, for his support of the minerals outreach program and for funding of this project; Stu Carlson, the minerals outreach coordinator who advocated funding the project; John Santora, the BLM-Alaska Deputy State Director for Mineral Resources, for his support of the project; John Cook, Roger Delaney, Cal Westcott, Sandy Westcott, Don Knight, and Donald Keill for their contributions and advice; Jane Mangus and Marvin Mangus for their comments and contributions; Doug Lalla and Perry Francis, of the Branch of Pipeline Monitoring, for logistical support and Alyeska Pipeline Service Company, for access to their archives and logistical support.

A very special thanks to Jan Burris of the Arctic District Office for her ideas, enthusiasm and hard work in arranging logistical support; Jim Sisk of the Arctic District Office for his enthusiastic support of the geology field research; and geologist Gil Mull of the State of Alaska Division of Geological and Geophysical Surveys, for his review, contributions, comments and enthusiastic support.

The BLM Minerals Outreach Program was created by Hillary Oden, the Bureau of Land Management Assistant Director for Energy and Minerals Resources, and Stu Carlson. Minerals Outreach is a program designed to inform and educate the public about their public mineral resources and the professionals who manage them. The BLM-Alaska Division of Mineral Resources participates in the outreach program through books, scientific publications, and talks to schoolchildren. For more information concerning BLM-Alaska minerals publications, please contact :

Bureau of Land Management
Alaska State Office
Office of External Affairs
222 W. 7th Avenue, #13
Anchorage, AK 99513

Riches from the Earth

Traveling the Dalton Highway

Your trip up the Dalton Highway will take you through some of the most remote and spectacular land in Alaska. Most of this land is undeveloped and contains a wealth of unusual scenery, seldon-seen wildlife and recreational opportunities. The Bureau of Land Management manages about 2.7 million acres along the highway between the Yukon River and Toolik Lake. Much of the remaining land is owned by the State of Alaska.

To enjoy your trip the most, you **must** come prepared to face the unexpected.

Be aware that the Dalton is a rough gravel road whose main purpose is a supply route for the North Slope oil fields. Be prepared to encounter fast-moving tractor-trailer rigs. Flying gravel from passing vehicles is a significant road hazard. Travel delays can be encountered during the summer due to road maintenance and repair.

The road can be very dusty or slippery, depending on the weather. Drive with your headlights on at all times. If you must stop, pull as far as possible to the right as you safely can, or park on waysides or access roads (do not block gates).

Off-road vehicle use is restricted along the Dalton Highway.

Make sure your vehicle is in excellent working order. Check your spare tire. Carry extra food, water and clothing.

Commercial facilities are extremely limited. Gasoline, food, lodging, a telephone, tire repair and emergency towing are available only at the Yukon River Crossing, Coldfoot and Deadhorse. Pipeline pump stations cannot provide assistance to travelers.

Informal camping is the rule along the Dalton but the Bureau of Land Management is gradually developing waysides and camping areas. Currently the only developed campground is scheduled for completion at Marion Creek, mile 180, in 1994. BLM operates an information station at the Yukon River crossing and an interagency visitor center at Coldfoot during the summer months.

For further information about traveling the Dalton Highway, contact:

Bureau of Land Management
Arctic District Office
1150 University Avenue
Fairbanks, Alaska 99709-3844
(907) 474-2301

Note: As of this writing, the State of Alaska has restricted travel in private vehicles past Disaster Creek (mile 211). Contact the Alaska Department of Transportation at (907) 451-2200 for the latest information.

Foreword

Riches—more than you can imagine! Welcome to the Dalton Highway, arguably the most unusual highway in the United States. This highway may best illustrate how the historic quest for mineral resources opened the nation. It owes its existence to oil and gas development on Alaska's North Slope, which currently supplies 25 percent of our nation's petroleum production.

Oil and natural gas are not, however, the only treasures tapped by this roadway. You are about to embark on an adventure into Alaska's arctic region. As you travel, you will cross the legendary Yukon River, the Arctic Circle, the Brooks Range, and the vast plain of the arctic tundra. You will pass through areas full of frontier mining lore, where miners toiled against nature hoping to carry away golden riches. You will cross areas where early prospectors, surveyors and geologists struggled to study a wilderness, and in the process, discovered a new gold—oil.

Few people today can imagine the rigors these early explorers endured: winter temperatures to 70 degrees below zero; living off the land; relentless swarms of mosquitos; no human contact for months at a time; no roads and few trails. The names they left behind commemorate their struggle: Noluck Lake; Disappointment Creek; Desperation Lake. By enduring their hardships they have unveiled "the last frontier."

This guidebook is written for the layman with an interest in geology. It explains the general geologic processes which created the features and scenery along the roadway. It also points out historical and general information. Where possible, we have simplified geologic concepts. We have emphasized specific geologic terms in bold type and define them in a glossary at the back of the book.

As you view the changing scenery, try to imagine the tremendous geologic forces that created and destroyed ancient tropical oceans and mountains, and formed the Brooks Range and the giant oil fields of the North Slope. Picture the vast glaciers that carved the valleys through which caribou migrate. Imagine the geologic changes that transformed a land where dinosaurs once roamed to a land where caribou now roam.

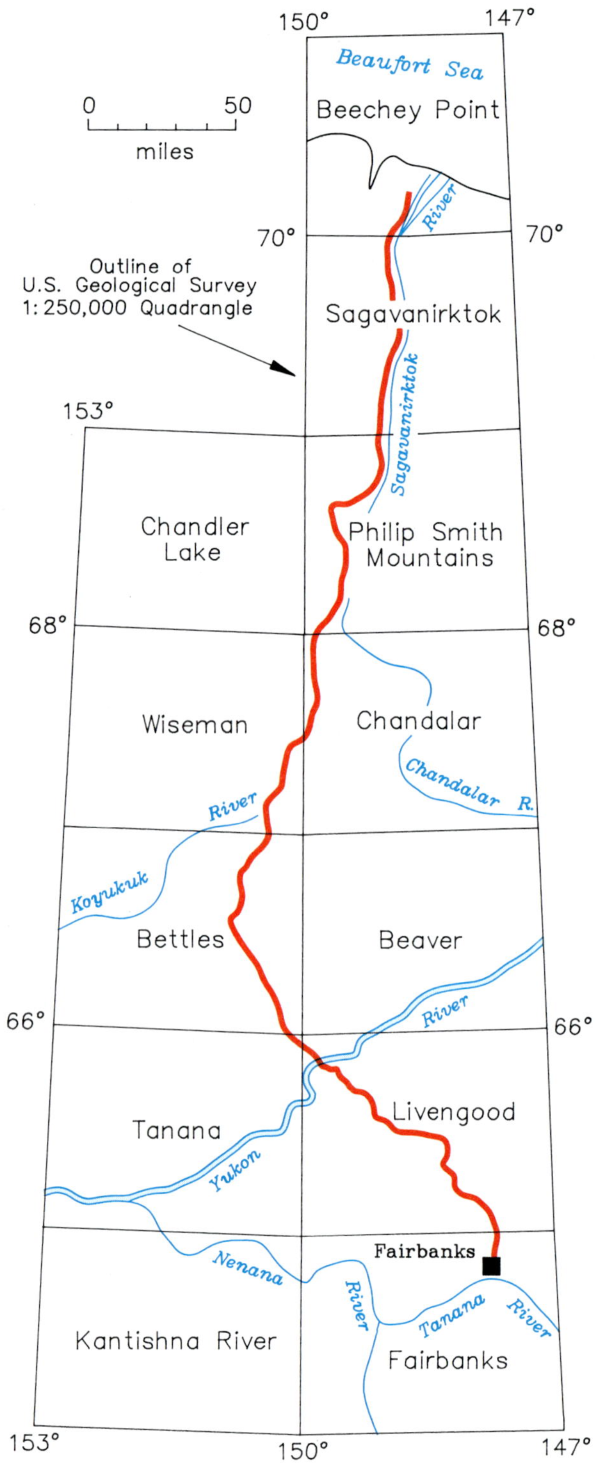

NOTE: U. S. Geological Survey topographic maps are available at U. S. Government public inquiries offices as well as many public map sales outlets. Prices are nominal. The maps identified by the name shown at the left (eg. Chandalar) are at a scale of 1:250,000 or 1 inch = 4 miles. More detailed maps at a scale of 1:63,360 or 1 inch = 1 mile are also available.

Figure 1. Index to 1:250,000 scale U. S. Geological Survey topographic maps covering the guidebook route.

INTRODUCTION

Oil — Catalyst for Change

The discovery of North America's largest oil field, Prudhoe Bay, in 1968, sparked events that would change Alaska's economic future for decades to come. An estimated 10 billion barrels of recoverable oil (one barrel equals 42 gallons) and 26 trillion cubic feet of natural gas lay 8,000 to 10,000 feet below the surface—in oil industry terms, a "supergiant" among oil fields.

A flurry of activity followed the Prudhoe Bay discovery. In anticipation of future lease sales by the State of Alaska, oil companies scrambled to collect additional geologic and geophysical data. This information, when combined and interpreted, portrays a three dimensional geologic model of the Prudhoe Bay subsurface.

By 1969, plans were drawn to build an 800-mile pipeline from the oil field at Prudhoe Bay to the ice-free port at Valdez. Alternative methods of oil transportation were considered, including oil tanker routes from Prudhoe Bay to the U.S. east coast via the Northwest Passage. The successful voyage, in 1969, of the *USS Manhattan* through this passage proved the alternative feasible but not as practical as the pipeline.

In 1970, a consortium of eight major oil companies formed Alyeska Pipeline Service Company. Alyeska undertook the monumental task of designing, building, and operating a 48-inch-diameter pipeline capable of transporting two million barrels of oil a day, 365 days a year.

On November 16, 1973, after a lengthy battle over environmental concerns, Congress authorized pipeline construction. The legislation passed by a margin of one vote. Vice-President Agnew cast the tie-breaking vote. This authorization closely followed the 1973 Arab oil embargo.

The initial cost of the Trans-Alaska Pipeline System, or TAPS, was estimated at about $900 million. However, actual dollars spent on pipeline construction, including workpads, nine pump stations, and the marine terminal at Valdez totaled about $8 billion. The pipeline construction work force, which at its peak consisted of more than 28,000 employees and contractors, spent a little over three years building a pipeline that crosses 800 rivers and streams (including the Yukon River), three mountain ranges, and is designed to withstand an earthquake of up to 8.5 on the Richter Scale.

Original pipeline design called for 12 pump stations, but the use of a drag reduction agent, which reduces the turbulence in the oil, allows only 10 operating stations to pump the pipeline's maximum of 2.136 million barrels per day.

The first North Slope oil entered the trans-Alaska pipeline in June of 1977. In 1988, the pipeline averaged over 2 million barrels of oil per day. By the end 1992, over 9 billion barrels of oil had passed through the terminal at Valdez. More than 12,000 tanker trips have carried North Slope oil to ports along the west coast of the U.S.

The Haul Road—Highway to the Arctic

Before the first pipe could be laid along the northernmost route of the pipeline corridor, a road had to be built—one extending north 358 rugged miles from the Yukon River over the Brooks Range to Prudhoe Bay (fig. 1). This huge undertaking, considered a vital link in the overall effort, began in the spring of 1974 and ended just 154 days later. It was known as the "Haul Road" and it became the lifeline that ferried men, women, machines, and material over rivers and across the **permafrost**-supported tundra to points along the northern pipeline route.

In 1981, the Alaska State Legislature formally named the Haul Road, and the 56 mile road extending from near Livengood to the Yukon River, after James Dalton. Dalton was an early Alaskan entrepreneur as well as an engineering consultant and construction expert for some of Alaska's early oil exploration efforts.

Origins: The Geologic Setting

With the widespread acceptance of the **plate tectonics** theory, many geological interpretations now include **terrane** descriptions. Terranes are **fault**-bounded groups of rocks that differ significantly in rock types or structures, or both, from the surrounding rocks. Each terrane has its own unique geologic history. Ideally, terranes represent the identification of features such as **island arcs**, ocean floor or continental fragments amalgamated or "welded" onto a continent through tectonic processes. (Envision an island archipelago such as the Philippines or Cuba eventually becoming part of the nearby continental landmass.) Terranes also result from displacement of one part of a continent to another location via **strike-slip faults**. Such is the case of the land west of the San Andreas Fault in southern California.

Terranes represent some of the most dynamic and exciting aspects of plate tectonics (fig. 2). Some geologists suggest that about 50 separate terranes comprise Alaska (fig. 3). Some geologists suggest fewer. Still others do not accept the terrane idea at all!

About 200 hundred million years ago (according to the terrane theory), the first Alaska terrane, the Yukon-Tanana terrane, collided with the northwesternmost margin of the North American **craton**. By late Mesozoic time (about 70 million years ago), most of these colossal crustal blocks, which represent diverse geologic environments, had assembled into what was to become Alaska.

Interpretations by scientists of **paleomagnetism**, as well as the fossil record, reveal startling evidence in support of the theory that some terranes may have traveled hundreds and even thousands of miles before docking in Alaska.

Terrane interpretations are in a state of evolution because many geologists do not agree on the sizes and boundaries of all the terranes. However, by gaining an understanding of terrane origin, of terrane movement, and of terrane welding onto the continental framework, geologists can more readily identify and evaluate areas favorable for the occurrence of specific types of mineral deposits.

Figure 4 shows the general extent of the major terranes in north-central Alaska. Each terrane is distinguished from adjacent ones by its internal geologic similarity. In some cases, abrupt changes in adjacent topographic features can help identify the boundary between two terranes.

Upwelling (A) of magma at a spreading center causes ocean crust/plate to move away from center. Where another plate resists or opposes this motion (B), mountains are heaved up (as in the Brooks Range), resulting in extensive faulting.

Subduction occurs where the oceanic crust dips under another plate (C) and is heated by pressure and friction, partially melting into magma. Magma rises to form volcanos (D) (as in the Aleutian Islands), and the remaining plate dips to be recirculated deep within the earth.

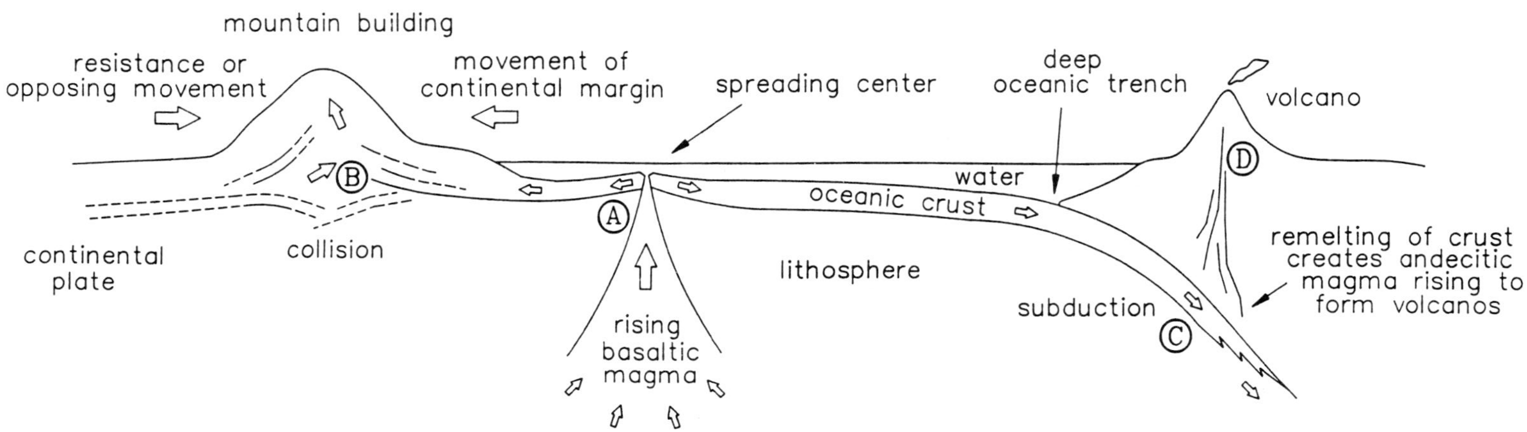

Figure 2. Plate tectonic process.

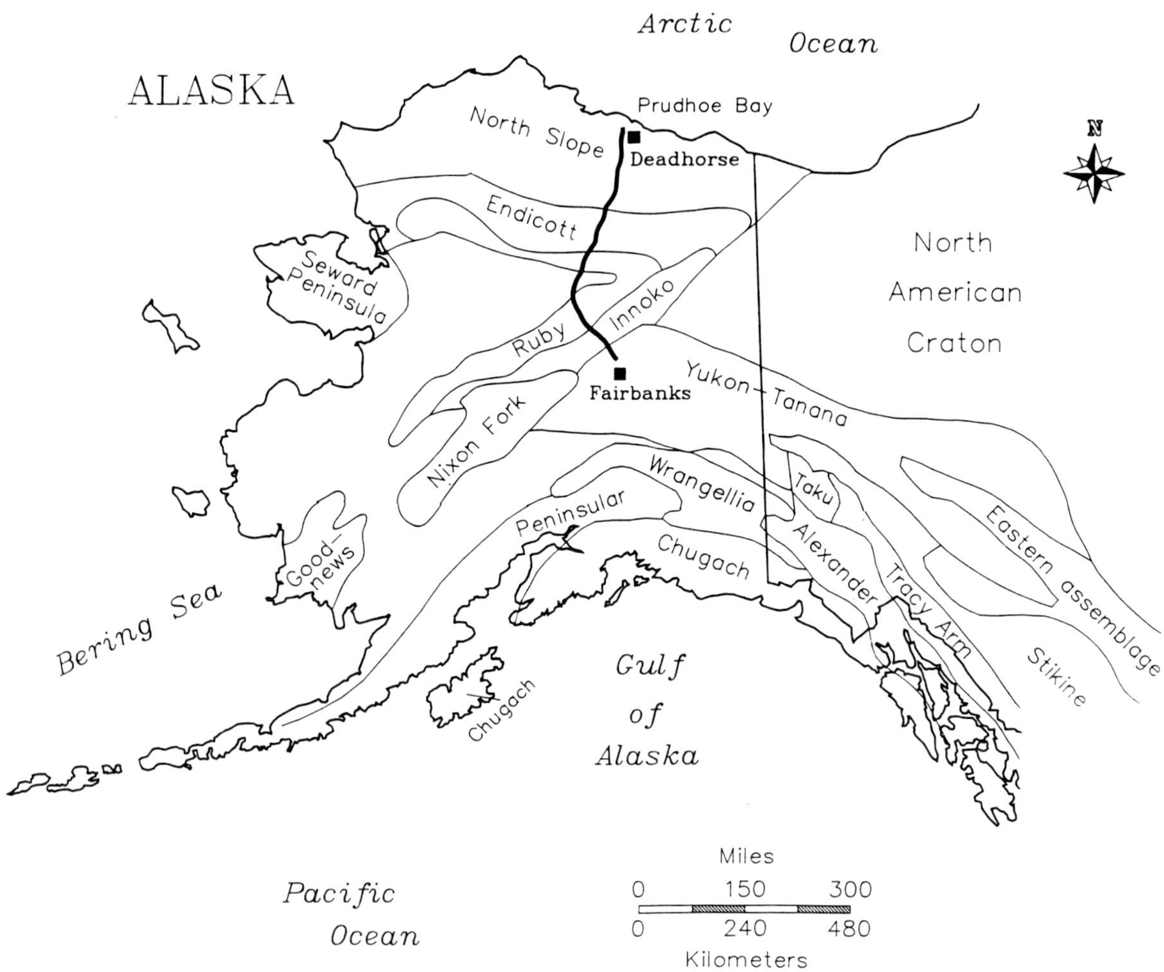

Figure 3. Generalized distribution of selected terranes in Alaska.
(after Jones and others, 1984)

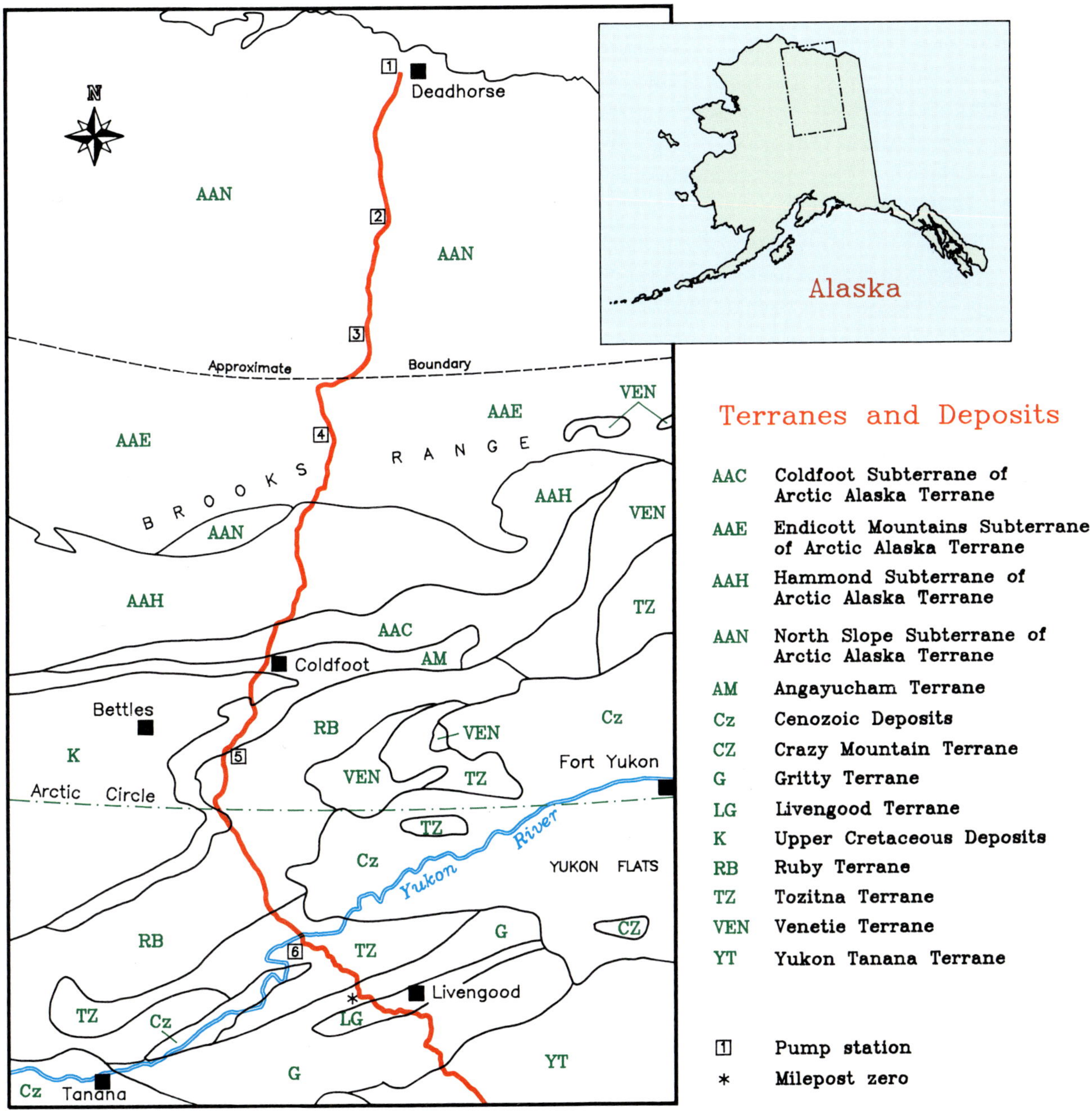

Figure 4. Terranes along the Dalton Highway.
(modified from Silberling and others, 1984)

Glaciers Shape the Land

Millions of years after compressional tectonic forces raised the Brooks Range above the surrounding lowlands, glaciers carved much of the spectacular countryside we see along the Dalton Highway. Major glacial episodes began about three million years ago during the middle to late Pliocene age and lasted until about 10,000 years ago.

Three principal intervals of glaciation have been defined through mapping of the surficial geology in the central Brooks Range (fig. 5). These intervals, separated by long periods of uplift and erosion, produced many of the surficial deposits within and along the northern and southern foothills of the Brooks Range.

Contrary to what you may expect, as much as half of Alaska remained free of glacial ice during these periods. Geological studies conclude that thick mantles of ice covered an area extending from the Alaska Range southward beyond Alaska's coast. Likewise, glaciers locked most of the central and eastern Brooks Range in thick mantles of ice. However, the land between these two great mountain ranges (which includes the first 120 miles of the Dalton Highway) and the land north of the Brooks Range, remained remarkably free of ice (fig. 6).

The explanation for this is that the great bulk of the Alaska Range induced precipitation from the moisture-laden air masses moving northward from the Pacific Ocean. This concentration of moisture nourished enormous ice sheets on the south side of the Alaska Range and denied the moisture needed for glacial formation in central Alaska and on the North Slope; only the higher elevations of the Brooks Range received enough moisture to nourish glacier formation.

The Brooks Range developed alpine glaciers (fig. 7) similar to those that formed in the Rocky Mountains of the western U.S.; these differed from the great continental ice sheets that covered much of central and eastern Canada and midwestern and northeastern U.S. Alpine glaciers originate as relatively

Figure 5. Central Brooks Range glacial sequence with suggested age assignments.
(modified from Hamilton, 1982 and Hamilton and Hopkins, 1982)

FAN MOUNTAIN	Late Holocene
ITKILLIK II	Late Wisconsinan
ITKILLIK I	Early Wisconsinan (?)
SAGAVANIRKTOK RIVER	Middle Pleistocene
ANAKTUVUK RIVER	Early Pleistocene
GUNSITE MOUNTAIN erratics	Late Tertiary

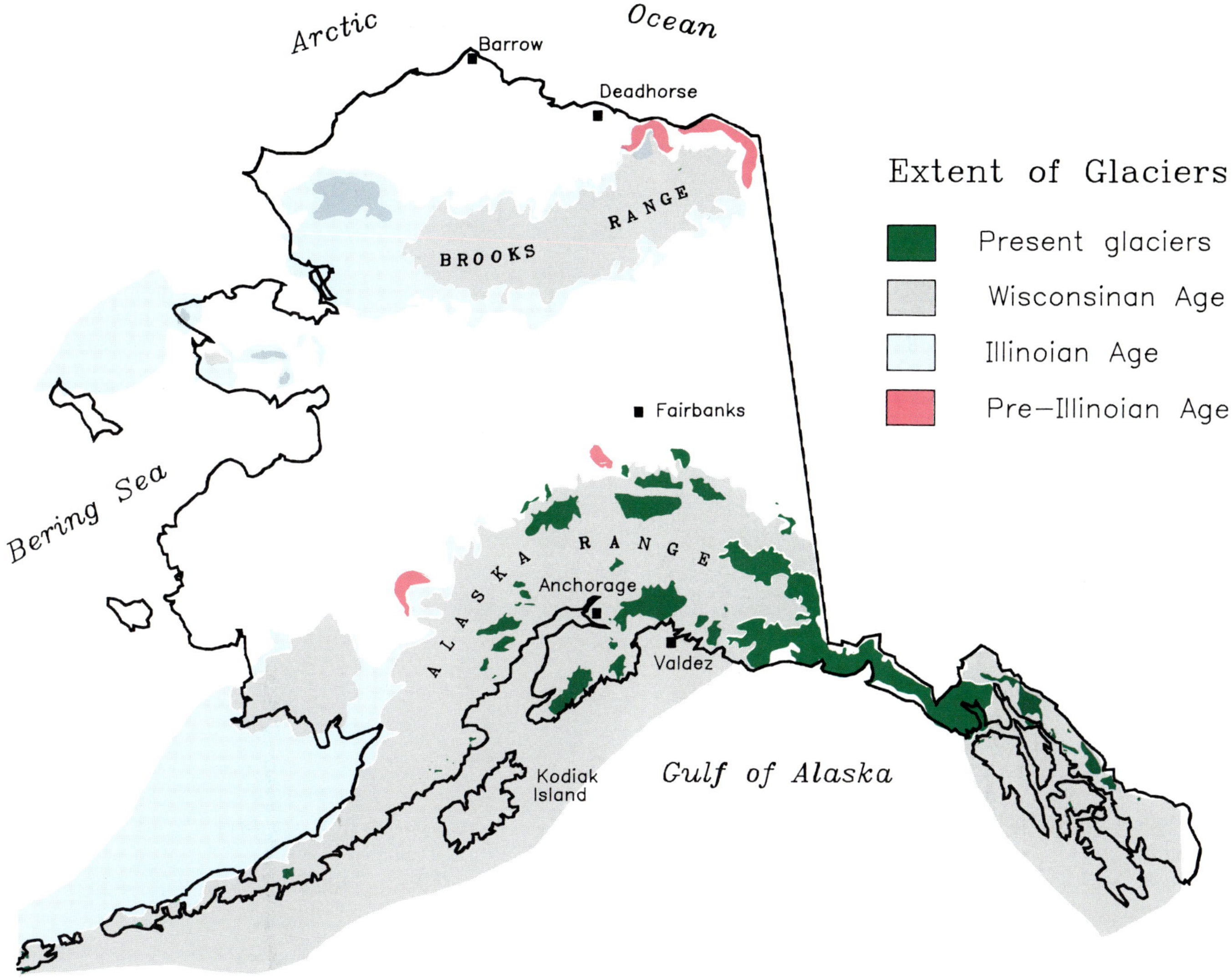

Figure 6. Extent of Quaternary glaciations in Alaska.
(after Coulter and others, 1965)

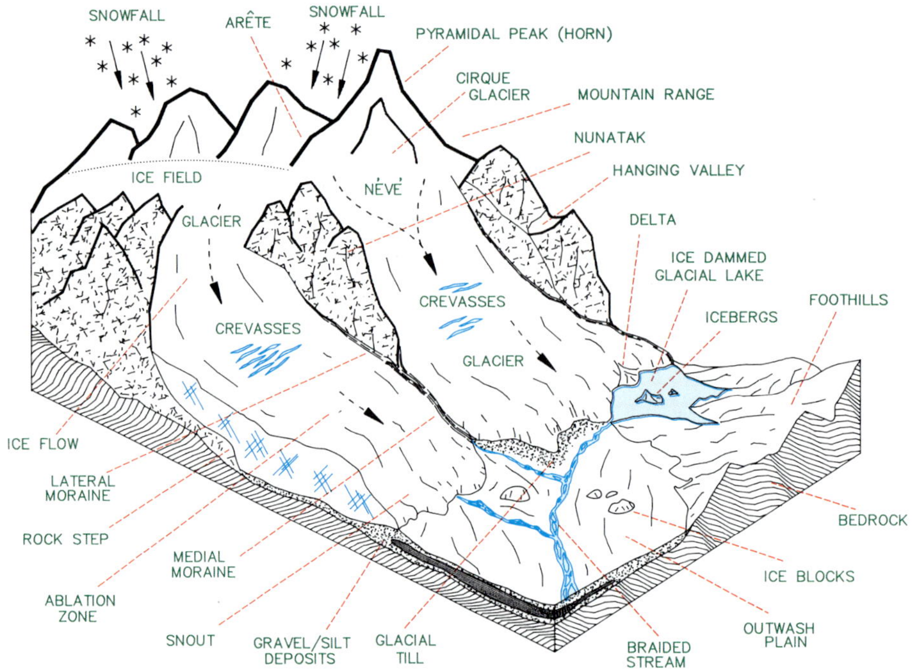

Figure 7. Glacial processes
(modified from Strahler, 1963)

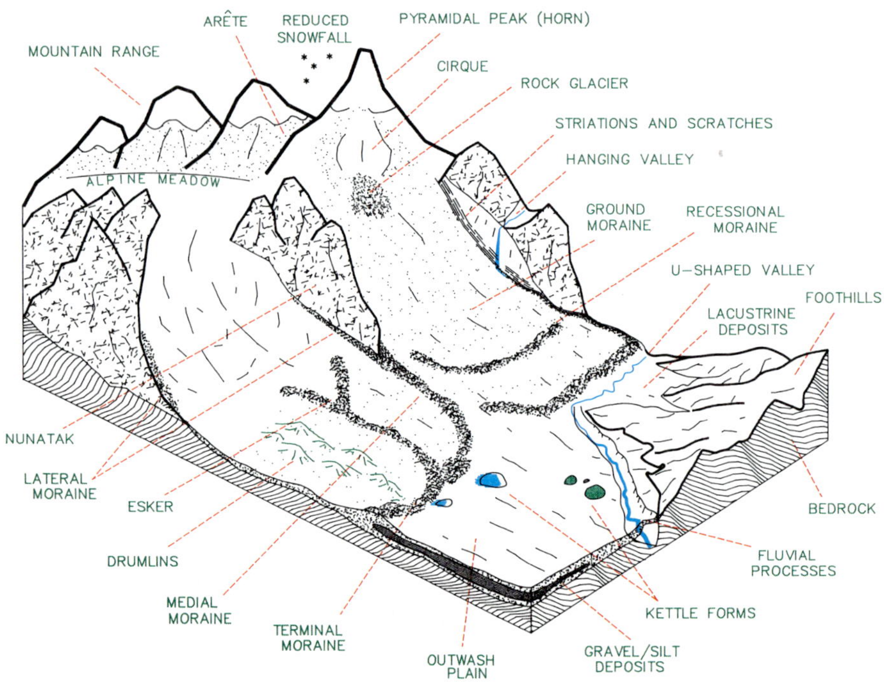

Figure 8. Post-glacial land features
(modified from Strahler, 1963)

small **cirque glaciers** high on the sides of mountains. (Several cirque glaciers are visible from the Dalton Highway between mileposts 272 and 280.) As cirque glaciers accumulate more and more snow, they begin to grow, flow downhill, and merge to become larger **valley glaciers**. Figure 8 represents land features left behind by glacial scouring and its subsequent melting and retreat.

During major glacial intervals of the Pleistocene Epoch, ice was sufficiently thick to cover most of the lower peaks of the Brooks Range and many of the ridges of the southern foothills. Glacial ice advanced northward and united across divides to form a series of piedmont lobes, some extending as much as 45 miles north of the Brooks Range. Despite the outpouring of glacial ice onto the foothills province, no more than 15 percent of the North Slope was ever covered with ice.

Today, Brooks Range glaciers decrease in number and size from east to west and are found mostly on north-facing, often oversized cirques in the higher elevations.

Glaciation and Wildlife Distribution

Toward the end of Pleistocene time, about 10,000 years ago, many mammals became extinct in Alaska, due in part to a limited ideal habitat and, in the case of grazing animals, increased predation by man. The most common and largest mammals were the mammoth, bison, and horse.

Climatic fluctuations since then have caused radical changes in the distribution of numerous plants and animals throughout Alaska. Today, the Dalton Highway corridor is home to a variety of wildlife including, but not limited to moose, caribou, beaver, lynx, fox, grizzly bear, bald eagle, peregrine falcon, trumpeter swan, grayling, and lake trout.

Permafrost and the Periglacial Environment

Permafrost, or perennially frozen ground, is present throughout 82 percent of Alaska and may occur in any material, even bedrock. Defined solely on the basis of temperature, permafrost forms when more heat leaves the ground than enters, and when the temperature remains below 32 degrees F for two years or more.

Several factors control the distribution of permafrost: climate, geology, topography, hydrology, and vegetation. **Ground ice** is the most prominent feature in permafrost that affects engineering activity in the north. It ranges from pore-fillings in sediments and bedrock to large ice lenses and wedges. Most ground ice is found in fine-grained, loosely arranged sediment and has a highly variable distribution.

Because permafrost and associated ground ice underlie much of the Dalton Highway, special engineering methods were adopted to ensure proper road design and to facilitate route selection, construction, and maintenance. Previous attempts at road construction in arctic Alaska had demonstrated that a delicate thermal equilibrium exists between the surface and the underlying permanently frozen ground. Destroying or removing the vegetative mat at the surface can disrupt this balance and induce permafrost thawing (Fig. 9).

Thawing permafrost produces drainage problems and sediment instability that can cause the ground to settle, cave in, or sink. In addition, excessive water within the **active layer** (the ground above permafrost which thaws in summer and freezes in winter) can cause ground surfaces, such as roadbeds, to heave upon freezing. To help maintain the thermal equilibrium, highway design called for a thick gravel pad to insulate the perma-

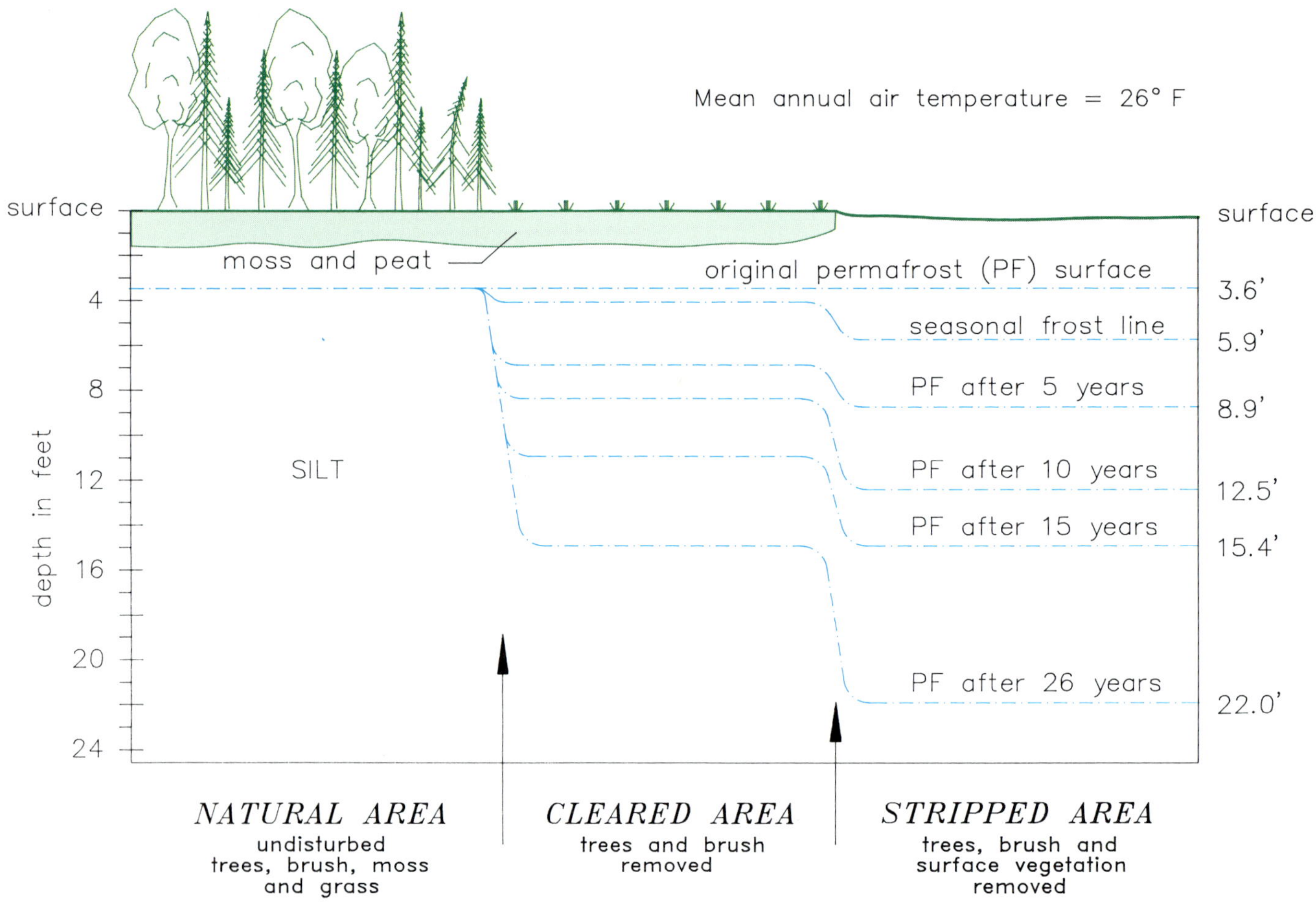

Figure 9. Permafrost degradation near Fairbanks, Alaska
(after Linnell, 1973)

frost and prevent it from thawing.

The rigorous climatic conditions that produce permafrost also generate other geologic processes that continually modify the landscape below the snowline, the **periglacial** zone. Within the periglacial environment, repeated cycles of freezing and thawing produce erosional debris that migrates downslope (under the influence of gravity) to form irregular landforms composed of loose rock material. Examples of resultant periglacial landforms, all visible from the Dalton Highway, include **patterned ground**, **pingos**, **rubble sheets**, and **rock glaciers**.

Dalton Highway Road Log

The road log is presented in a mile-post format. That is, the mileages used in this book correspond to the small, green mile markers posted along the Dalton Highway. In some places mile markers may be absent due to recent road maintenance or realignment. We, therefore, point out stream crossings and other useful topographic features to help approximate absent mile marker locations.

Mile 0

Intersection of the Dalton and Elliott highways. The first 56 miles of the Dalton Highway, originally called the "TAPS Road", were built in 1969 and 1970, shortly after the announced discovery of the Prudhoe Bay oil field in 1968. However, road construction north of the Yukon River was delayed several years by numerous court injunctions against the building of the pipeline. This road was eventually completed in July 1974.

Completion of the Dalton Highway provided access to north-central Alaska. Access to areas close to the highway, however, is still fairly difficult, time consuming, and expensive under the best of conditions. Navigable rivers do not reach all segments of this remote region, and helicopter time can exceed $500/hr! Consequently, much of the presently available geological mapping and interpretation is of a reconnaissance nature rather than detailed. (This changes yearly with additional work.) Thus, many tectonic, lithologic, and stratigraphic relationships are not well known or are difficult to reconcile as parts of a single geologic domain.

Geological mapping has revealed many different bedrock types across northern Alaska. Unfortunately, bedrock exposures, from which geologic history can be interpreted, are few and far between throughout much of central Alaska. This is due primarily to the lack of glacial sculpturing, which tends to remove accumulated surface debris and expose large areas of bedrock. What has developed instead

Refer to figure 10 for geology and points of interest between Mile 0 and Mile 80.

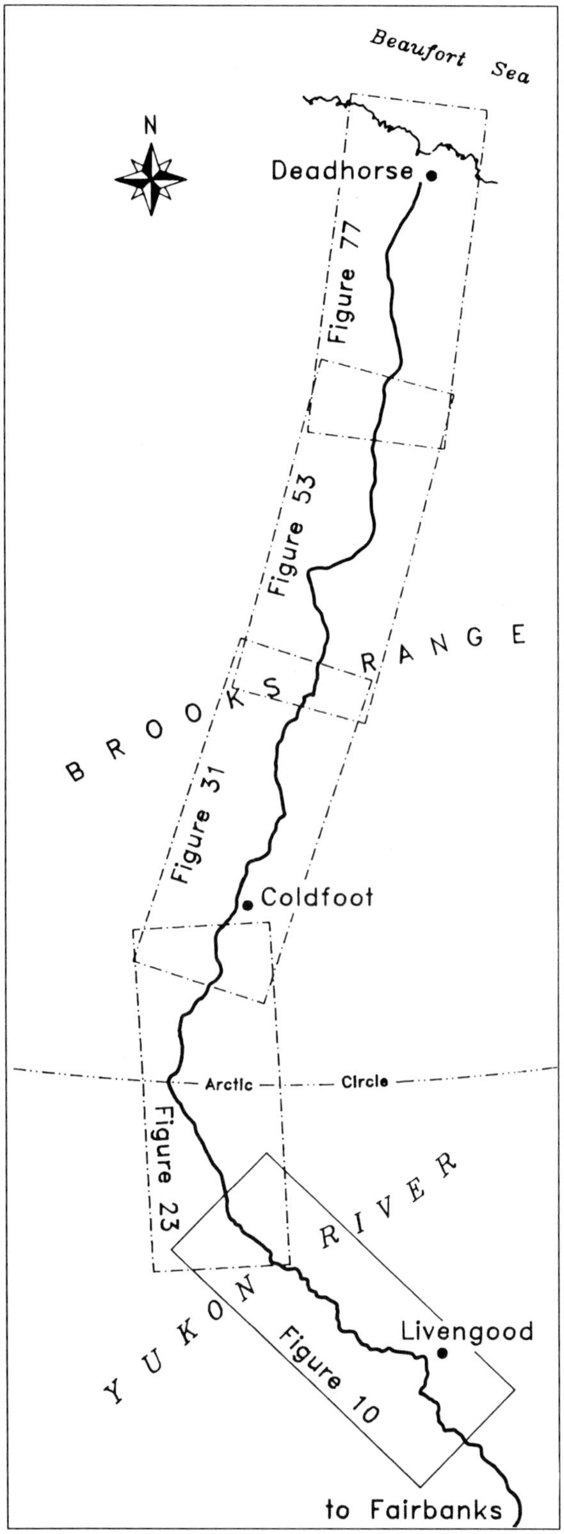

are thick and widespread periglacial deposits such as **loess**, **solifluction** deposits, and **colluvium** which conceals the underlying rock.

Much of the bedrock that is exposed in the areas between these deposits typically weathers to **frost-riven** rubble outcrops showing few of the local bedrock's distinguishing characteristics. In addition, some rock types, such as structurally weak shales and volcanic **basalts**, erode to sloping, vegetation-covered surfaces, or form rolling hills rather than cliffs.

Mile 0.5

Bedrock near Livengood consists mostly of **Ordovician** to **Devonian dolomite**, **limestone**, **silicified carbonate** rocks and **chert**. The road to the west leads to a rock quarry exposing interbedded **phyllite** and black chert, with silicified carbonates. About 380 million years ago the sediments which formed these rocks were deposited in a deep marine environment. Deep burial and tectonic pressures transformed (metamorphosed) the **sedimentary rocks** into metamorphic cherts and phyllites.

The quarry face also contains examples of **fault gouge** and **slickensides** which indicate that the bedrock was later faulted and **folded**.

A thin layer of wind-blown silt, called loess, covers the bedrock lying above the quarry face. Moderately thick deposits of loess cover most of the hills and ridges between here and the Yukon River. In this region, loess blows in from the extensive river and stream deposits of the Yukon Flats basin, a widespread lowland area about 50 miles to the northeast.

Loess, widely distributed in Alaska, originates in the vegetation-free flood plains of braided glacial rivers. Much of the silt blows far beyond its source and covers most areas lying below altitudes of 1,000-1,450 feet.

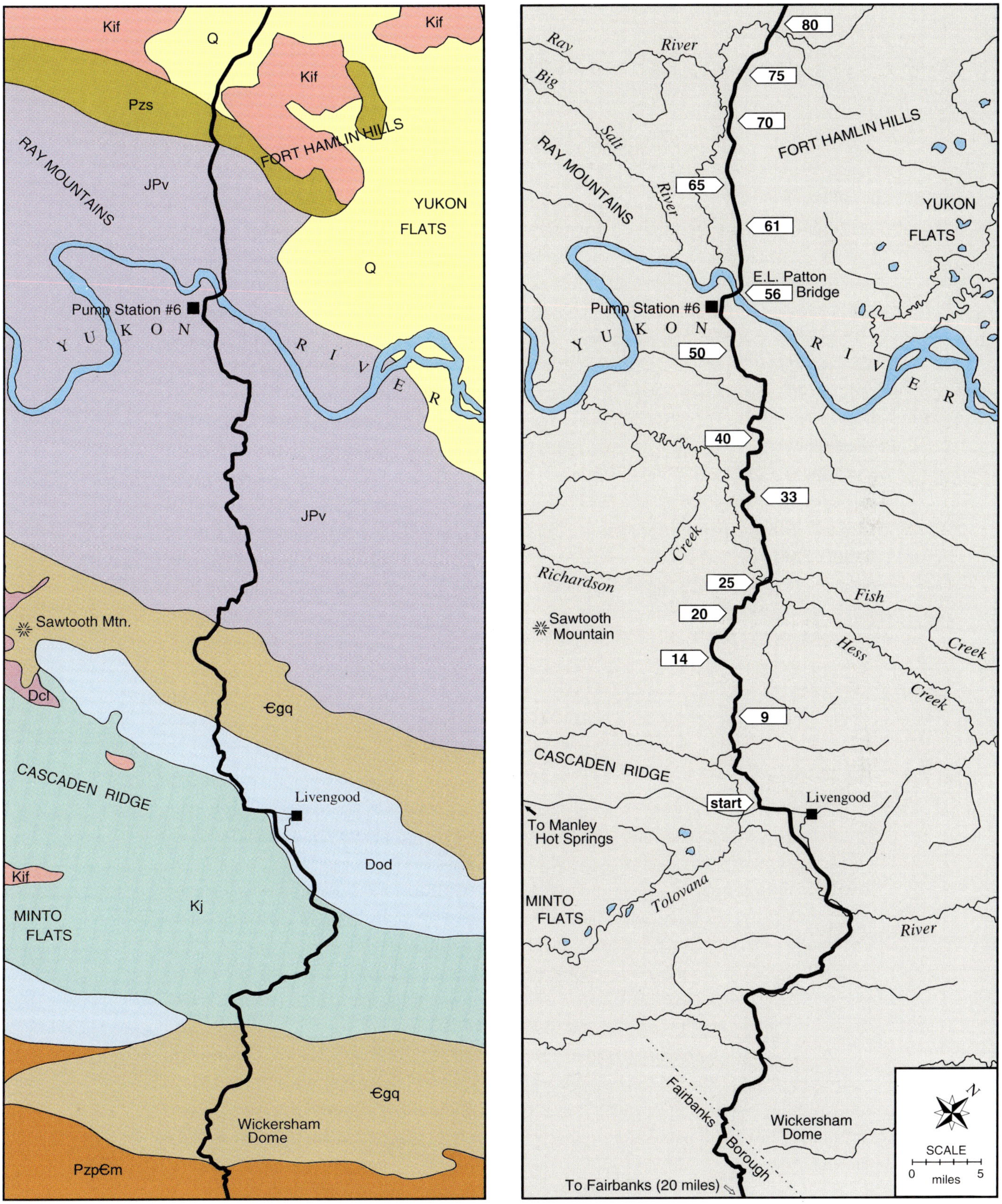

Figure 10. Dalton Highway geology and points of interest.
Mile 0 to Mile 80

(from Chapman and others (1971), Brosgé and others (1973), Patton and others (1973) and Beikman(1980). See figure 92 for key.)

YUKON-TANANA TERRANE

The Yukon-Tanana terrane (YT) is a regional terrane, that extends from central Alaska into the Yukon Territory (fig. 4). You will cross this terrane along the Elliot Highway on your way to the Dalton Highway. It extends about 35 miles north from the Fairbanks area to just north of Wickersham Dome. The relatively low, rounded hills consist mostly of light colored to silvery, **quartz**-mica **schists** and **gneiss** that have been deformed several times. Quartzite **veins** cut across all these rock types in striking fashion. These rock units form the Birch Creek Schist. They are of pre-Cambrian age and are the source of much of the gold that has been mined from the **placer** deposits in the Fairbanks area. **Granites intrude** some of these **metamorphic** rocks.

The Birch Creek Schist imposes a pronounced northeast trend upon the strike of the mapped rock units and the major faults of the Yukon-Tanana terrane. Many of the hills and drainages also have this pronounced northeast trend. This northeast fabric of the landforms and geology persists to about the Atigun Pass area (approximately mile 246).

"GRITTY" TERRANE

Immediately north of Wickersham Dome to about Livengood Creek, the Elliot Highway crosses mostly dark-colored, **shales**, **siltstones**, **sandstones** and **conglomerates**. Bedding is usually poorly developed, and most rock units are described as gritty. These unspectacular rocks have Paleozoic and Mesozoic ages, and they too have been intruded, locally, by granites of **Cretaceous** or **Tertiary** ages. Several terranes have been suggested for this area, but the nature of exposures makes it difficult to define and separate them. Due to limited space and for simplification, figure 4 simply refers to them as the "Gritty" terrane (G).

We are fortunate that there are accessible outcrops or road cuts along the Dalton Highway which expose some of the rocks within each terrane. In addition, the road log notes and explains some of the geologic processes, such as glacial abrasion, that have altered and obscured many of the bedrock exposures along the highway.

Alaska loess ranges in thickness from a fraction of an inch to more than 195 feet! It forms the soil of principal agricultural areas in Alaska and is, therefore, of great economic importance.

Over the centuries, large quantities of this wind-blown silt washed into valley bottoms to form thick deposits of perennially frozen silt—rich in organic sediment called muck. The muck deposits in Alaska have yielded partial carcasses of well preserved bison, mammoths, and horses, buried between 70,000 and 10,000 years ago (fig. 11).

Alyeska Pipeline Service Co.

Figure 11. Alaska's muck deposits (in background) yield relics from the last ice age such as this tip from a mammoth's tusk. The ivory tusk was recovered from a road construction site located about five miles north of the Yukon River.

Mile 6.1

An ancient **alluvial terrace** on the west side of the road lies 160-200 feet above the modern channel of Lost Creek.

Mile 14

The pipeline passes under the road. About half of the trans-Alaska pipeline is buried in areas where thawing would not cause loss of soil support for the pipe. These thaw-stable conditions exist in areas relatively free of excess ground ice or in areas containing no permafrost.

Where ground ice is abundant, the pipeline is elevated on vertical-support members or VSMs (fig. 12). About 78,000 VSMs (spaced 60 feet apart) support the elevated sections of the 800-mile-long pipeline. About 57 percent of the pipeline is above ground north of the Yukon River.

Heat retained in the active layer during the winter months is transferred from the soil to the VSM, where it rises, via heat pipes, through the supports to fin-like radiators made of aluminum. The heat pipes, which contain liquid ammonia and ammonia vapor, are designed to provide a continuous, nonmechanical heat removal system that prevents the thawing of permafrost around the VSM. In winter,

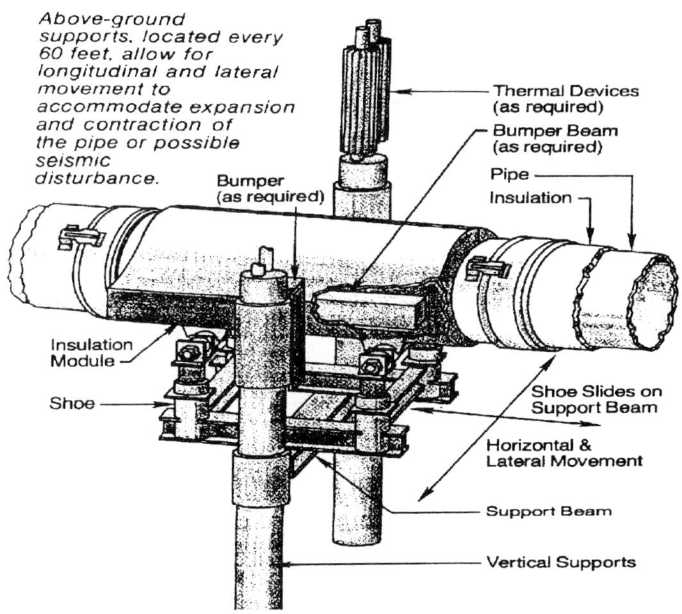

Figure 12. Vertical support member.

the cold ammonia vapor at the top of the heat pipe condenses and sinks to the bottom where it absorbs heat. The liquid ammonia turns to vapor and rises to the cooling fins which dissipate the heat.

Above-ground sections of pipeline were built in a flexible zigzag configuration to allow controlled sideways movement during earthquakes or during pipe expansion and contraction from heat or cold. The pipe, which measures 48 inches in diameter, has a wall thickness of either .462 or .562 inches and is constructed from high-tensile-strength carbon steel.

The rugged, bedded volcanic rocks and limestones of the White Mountains can be seen in the distance to the east. Their name is taken from the light-colored Tolovana Limestone which overlies and contrasts with the older, darker rocks of the Fossil Creek Volcanics. These limestone and volcanic rocks formed about 400 million years ago in a shallow, marine environment.

Mile 20

Look for the first of five major, deep, ice-rich road cuts excavated during construction of the first 56 miles of the Dalton Highway. Degradation of these cuts by thawing of the newly exposed ice-rich soil threatened roadbed stability (fig. 13). Highway engineers used near vertical cuts in the remaining four road cuts to expose less ice-rich soil to melting. This allows the vegetative cover to remain intact and slowly slump over to cover and insulate the receding surface (fig. 14).

Mile 20.3

The original ice-rich cut extended 30 feet above the road surface.

Figure 13. Deterioration of newly exposed ice-rich roadcuts along the construction route of the Dalton Highway compromised roadbed stability.

Alyeska Pipeline Service Co.

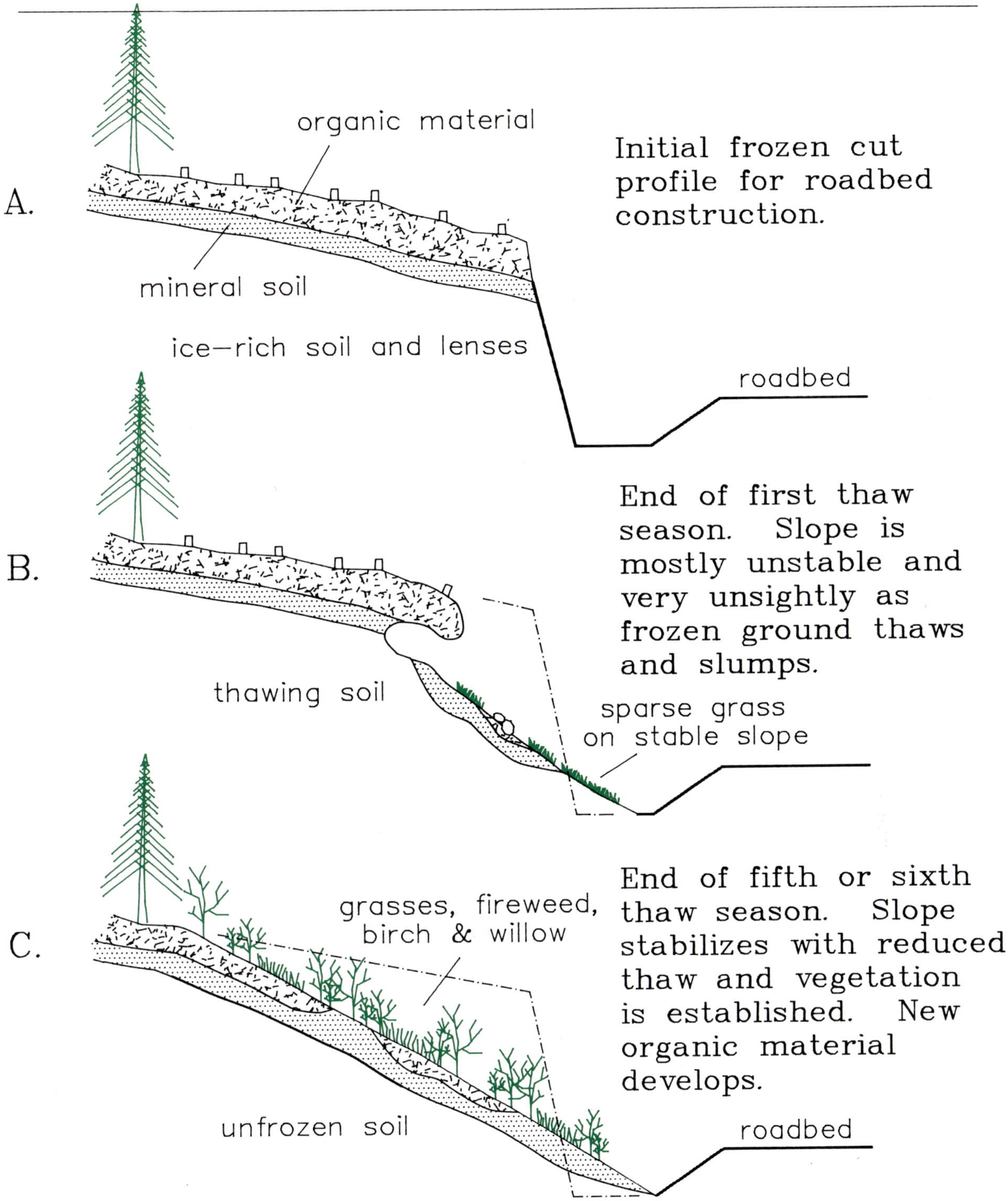

Figure 14. Development of stability in ice-rich road cut.
(after Brown and Berg, 1980)

Mile 20.9

Road cut through dark colored, massive, fine-grained mafic volcanic rocks. **Joints** within the bedrock are pronounced and there are a few cross-cutting quartz veins.

Figure 15. Thermo-erosional niching along Hess Creek. Wave action and spring breakup-flooding melt the ice-rich sediments along the outer banks and bluffs of Hess Creek. The resulting scalloped or overhanging banks eventually collapse exposing fresh surfaces to thermal erosion. Note the large ice wedge exposed in the frozen soil of the bluff.

Mile 23.8

Thermo-erosional niching along Hess Creek. Wave action and spring breakup-flooding melt the ice-rich sediments along the outer banks and bluffs of Hess Creek. The resulting scalloped or overhanging banks eventually collapse and expose fresh surfaces to thermal erosion. The collapsed banks form tilted soil blocks which may temporarily retard lateral erosion (fig. 15).

Mile 25

Note the abrupt change in vegetation between the outer and inner banks to the south along Hess Creek. The boggy area, laden with stunted black spruce, occupies the permafrost-rich outer bank. Vegetation, not tolerant of perennially frozen ground, colonizes and flourishes on the permafrost-free deposits of silt and gravel along the inside of the meander curve, or **point bar**. These point-bar deposits are left unfrozen as the river bend migrates westward thawing permafrost along its outer bank (figs. 16 and 17).

Eventually, in about 200-300 years, permafrost conditions along the inner bank will return as the insulating effect of the thick organic mat, produced by decaying vegetation, reduces the mean annual ground temperature once again.

Mile 29.9

A bedrock exposure east of the road is composed of light-colored, medium-grained granite that is severely **weathered**.

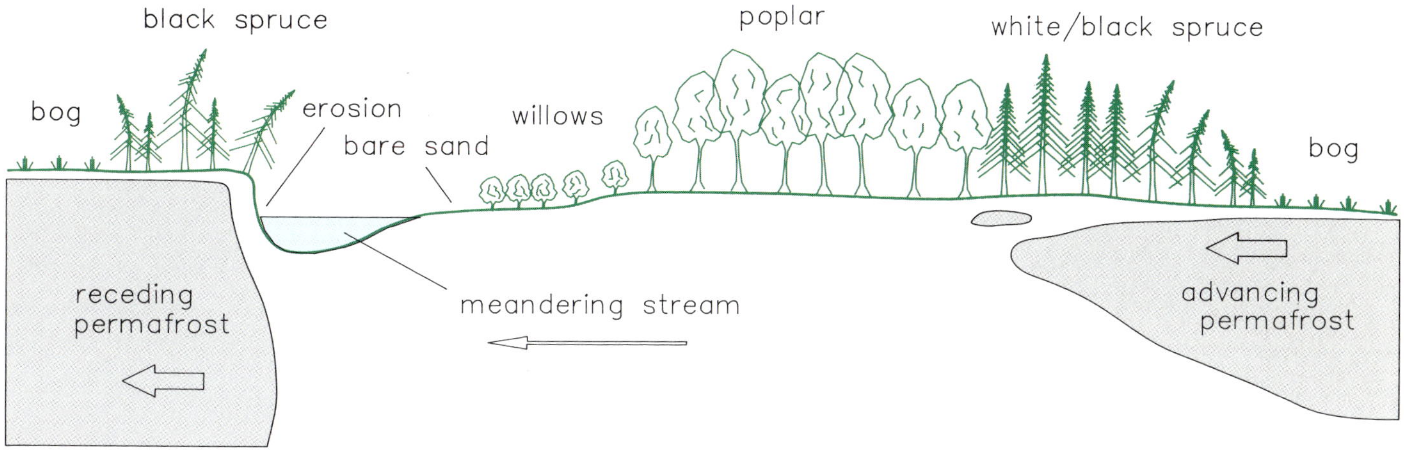

Figure 17. Effects of stream migration on permafrost and local vegetation
(after Kreig and Reger, 1982)

USDI BLM photo / William Diel

Figure 16. The boggy area laden with stunted black spruce along the outer bank of Hess Creek is underlain by permafrost-rich soil. Vegetation, not tolerant of perennially frozen ground, colonizes and flourishes on the permafrost-free deposits of silt and gravel along the inside of the meander curve, or point bar. These point-bar deposits are left unfrozen as the river bend migrates westward, thawing permafrost along the outer bank.

Tozitna Terrane

Tozitna terrane rocks underlie the highway from about mile 15 to mile 65, with scattered outliers to about mile 82. Most Tozitna terrane outcrops are overlain or draped by periglacial deposits and frozen ground. Where visible, these rocks have an overall southerly dip and are severely folded and faulted. Consequently, the sequence of events is not well known.

The Tozitna terrane consists predominantly of fine-grained, massive, **mafic** rocks with interbedded basalt, relatively thin beds of radiolarian chert, and minor amounts of limestone.

This petrologic assemblage and the presence of pillow-shaped lavas indicate that the Tozitna rocks formed in deep oceanic environments (fig. 2). These environments include mid-oceanic ridge geology, such as inter-plate spreading centers where crust is generated and drives plate tectonics. Thus, the Tozitna rocks represent remnants of the sea floor from the **Mississippian** through the **Triassic** age. In addition, a few outcrops have intrusions of **Jurassic gabbro**. Overall, these rocks were formed in high temperature regimes. These rocks become relatively unstable when exposed at the surface where they weather very easily. Thus, large and informative outcrops of Tozitna rocks are not common. Their areal extent is inferred, however, from magnetic and gravimetric geophysical surveys.

Mile 32.8

Pingos on west side of road. A mature spruce-birch forest grows on the top and flanks of these open-system pingos and contrasts with the scattered and deformed small spruce trees in the surrounding lowlands.

Pingos (Eskimo for "small hill") are conical ice-cored hills or mounds, 50 to 1,450 feet wide and 10 to 230 feet high that form when large massive ice layers grow near the surface in permafrost.

Two types of pingos are recognized in Alaska: the closed-system pingos of the Arctic (described at MP 377) and the smaller, open-system pingos of interior Alaska (fig. 18 and 19).

The open-system pingo usually forms on gentle to moderate slopes of **alluvium**-filled valleys where water beneath or within the permafrost penetrates the permafrost under high hydraulic pressure, then freezes. This pressure, along with water-to-ice expansion, heaves the overlying material into a mound. Almost all open-system pingos lie in unglaciated parts of the country. In Alaska, pingos are less than 7,000 years old.

Road cut on the east side of the road exposes dark-colored, volcanic rocks intruded by light-colored **diorite dikes**. The bedrock is extensively fractured, faulted, and contorted.

USDI BLM photo / Arthur Banet, Jr.

Figure 18. An open-system pingo (the small mound in the middle of the photo) has formed in this alluvium-filled valley. Water beneath or within the permafrost penetrates the permafrost under high hydraulic pressure, then freezes. This pressure, along with water-to-ice expansion, heaves the overlying material into a mound.

Mile 38

Road crosses over pipeline.

Mile 40

View of Sawtooth Mountains to southwest. A quarry at the top of the hill contains poor exposures of chert-pebble conglomerate of probable Cretaceous age. The angular, pebble-conglomerate is hard and compact, whereas the cobble conglomerate is **matrix** supported and relatively soft.

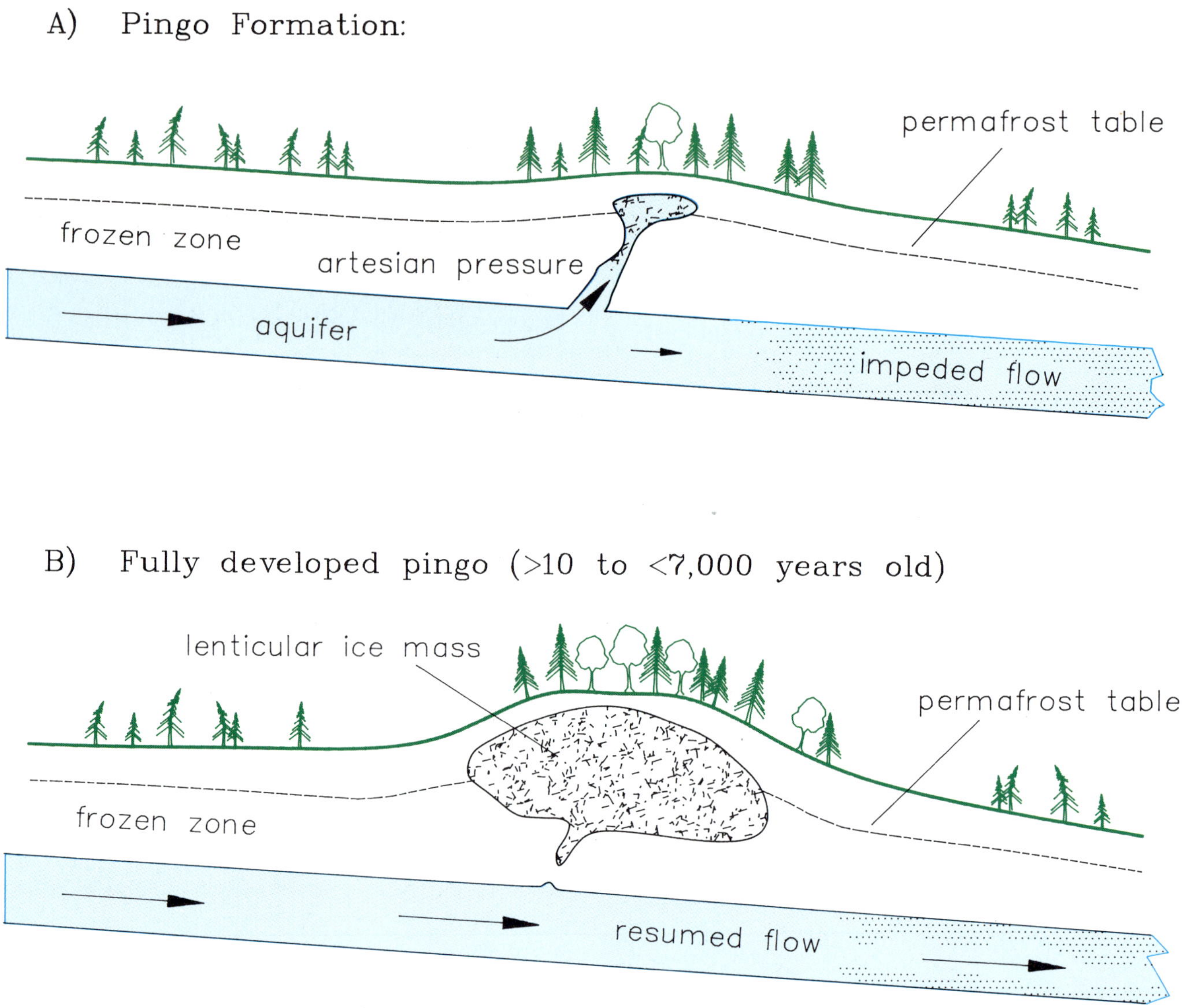

Figure 19. Open system pingo development.

Ruby-Arctic Alaska Terrane and the Granitic Plutons

The Ruby terrane (Rb) extends through a large area from the Yukon River to the south side of the Brooks Range. It consists of relatively low-temperature and low-grade metamorphic rocks. Along the highway, there are light-colored, shiny (when its sunny) phyllites and schists with some thin interbedded quartzose **graywacke** and phyllitic shale. Quartz veins are common, and, in places, have been deformed into sausage shapes (boudinage). The beds tend to be tightly folded, often into isoclinal folds where both limbs dip in the same direction. In places, the schists and shales have been moderately to highly metamorphosed.

Ruby terrane age estimates range from Proterozoic through Paleozoic. The rocks represent sedimentary processes similar to those now taking place off the eastern coast of the United States.

The rocks of the Ruby terrane form the core of a regional, northeast-trending, positive area called the Ruby **geanticline**. This part of Alaska may have remained above sealevel and may not have received any sediments since the mid-Cretaceous.

Outcrops of the metamorphic rocks of the Ruby terrane are seldom abundant or dramatic. But, they are more common than the sparse outcrops and roadcuts of the Tozitna terrane. Large bodies of granite (for example, Kanuti, Hot Springs, Jim River, and Bonanza Creek **plutons**) cut the terrane into discontinuous exposures. These plutons consist mainly of coarse-grained, **porphyritic** (having at least two distinct crystal sizes) biotite granites. (Large crystals of potassium **feldspars** can be found in the Finger Mountain area, MP 100.) The granitic plutons are thought to have formed from the melting and partial assimilation of ancestral Ruby terrane rocks within the earth's crust. Radiometric dating places the time of cooling of the melted rock between 104 and 112 million years ago.

Figure 20. The trans-Alaska pipeline crossing of the Yukon River was completed in 1976. The pipe was mounted on special supports on the side of the 2,700-foot long E. L. Patton bridge.

Mile 42

Road cut exposes weathered metamorphic rocks. There are red and green cherts that are now so contorted and sheared that this outcrop has no discernable top or bottom. Some of the chert even forms orbs.

Mile 50.5

Thermokarst pond east of road.

Mile 56

The Yukon River and E.L. Patton bridge. The Yukon River the largest river in Alaska, flows 1,979 miles from Canada to the Bering Sea. It is the 5th longest river in North America, and ranks 21st in the world (figs. 20 and 21).

The high bluffs along the Yukon River provide good exposures of the mafic volcanic rocks of the Tozitna terrane. River cobbles provide a good overview of the mafic rock types found in this area.

Mile 60.7

Road crosses over an elevated section of pipeline built on thaw-unstable soils as much as 30 feet deep.

Mile 61.8

Road to the east leads to Department of Transportation and Public Facilities (DOTPF) maintenance camp and a rock quarry that exposes gabbro of the Tozitna terrane. The Tozitna terrane consists of Mississippian to Triassic basalts and radiolarian chert, Permian limestone, and Jurassic **ultramafic** rocks intruded by **sill**-like gabbro bodies. Radiolarian chert, a compact variety of silica containing fossil shells of a microscopic, plankton-like marine animal, is exposed on the north side of the road just before the quarry entrance.

USDI BLM photo / Dewayne Richardson

Figure 21. View southeast of the volcanic rocks of the Tozitna terrane at the Yukon River crossing. Tozitna terrane rocks, which represent remnants of the ocean floor, formed between 380 and 208 million years ago.

USDI BLM photo / William Diel

Figure 22. Ice forming in the culvert under the road during winter months eventually blocks water that normally drains from the low spot on the east side of road. The rising water level within the low spot inundates the surrounding vegetation.

Upon freezing, ice attaches to the tree bark. As spring approaches, ice in the culvert begins to melt and the water level in the swale drops and causes the ice to collapse. This action can break trees near their base or simply strip them of their bark, thus exposing the wood to insects and disease, and eventually killing the tree.

Mile 65.5

On the right side of the road is a good example of how the road has altered the adjacent vegetation (fig. 22). Ice forming in the culvert under the road during winter months eventually blocks water that normally drains from the low spot on the east side of road. The rising water level within the low spot inundates the surrounding vegetation. Upon freezing, ice attaches to the tree bark. As spring approaches, ice in the culvert begins to melt and the water level in the swale drops and causes the ice to collapse. This action can break trees near their base or simply strip them of their bark, thus exposing the wood to insects and disease, and eventually killing the tree.

Mile 69.8-70.5

This road cut exposes a poorly preserved section of phyllite or scaly **argillite** and interbedded graywacke sandstone. These are possibly **turbidites** and are the southernmost exposures of the Ruby terrane along the Dalton Highway.

Mile 70

View of the Ray River. Another example, similar to Hess Creek, of the abrupt change in vegetation between the river's outer and inner banks—attributable to the thawing of soil as the outer bank of the river migrates over permafrost.

Mile 75

First views of granite tors (defined at MP 98.5) on Finger Mountain to the north.

Mile 75.5

Rubble sheets composed of Quaternary and Tertiary **olivine**-bearing basalts are exposed along both sides of the highway. The basalt has a vesicular nature, which means it

Angayucham Terrane

Rocks of the Tozitna terrane (Tz) and the Angayucham (Am) terrane are nearly identical. They differ mainly in geographic location. Angayucham terrane rocks are found from about mile 103 to about mile 168, where the highway intersects the south strand of the Angayucham fault. Through this area, Angayucham rocks commonly appear as widely scattered outliers atop, or in fault contact with, older rocks.

Current geologic thinking recognizes that these deep oceanic rocks are allocthonous (structurally out of place) as compared to the metamorphic rocks in this region.

Angayucham rocks along the north side of the Ruby terrane dip towards the north. As yet, there is no agreement as to how far these rocks have been moved, from what direction they have been **thrust faulted**, or how many times they have been deformed. Although most exposures of Tozitna and Angayucham rocks are thought to be separate fault blocks or slabs, current thought suggests that both the Tozitna terrane and the Angayucham terrane rocks (fig. 4) are pieces of a single, large allocthon. According to this theory, this allocthon was emplaced over the entire extent of the Ruby terrane. Tectonic reconstructions requiring the movement of such a large allocthon tax the creative imaginations of even the most pro-terrane-interpretation advocates!

Refer to figure 23 for geology and points of interest between Mile 80 and Mile 165.

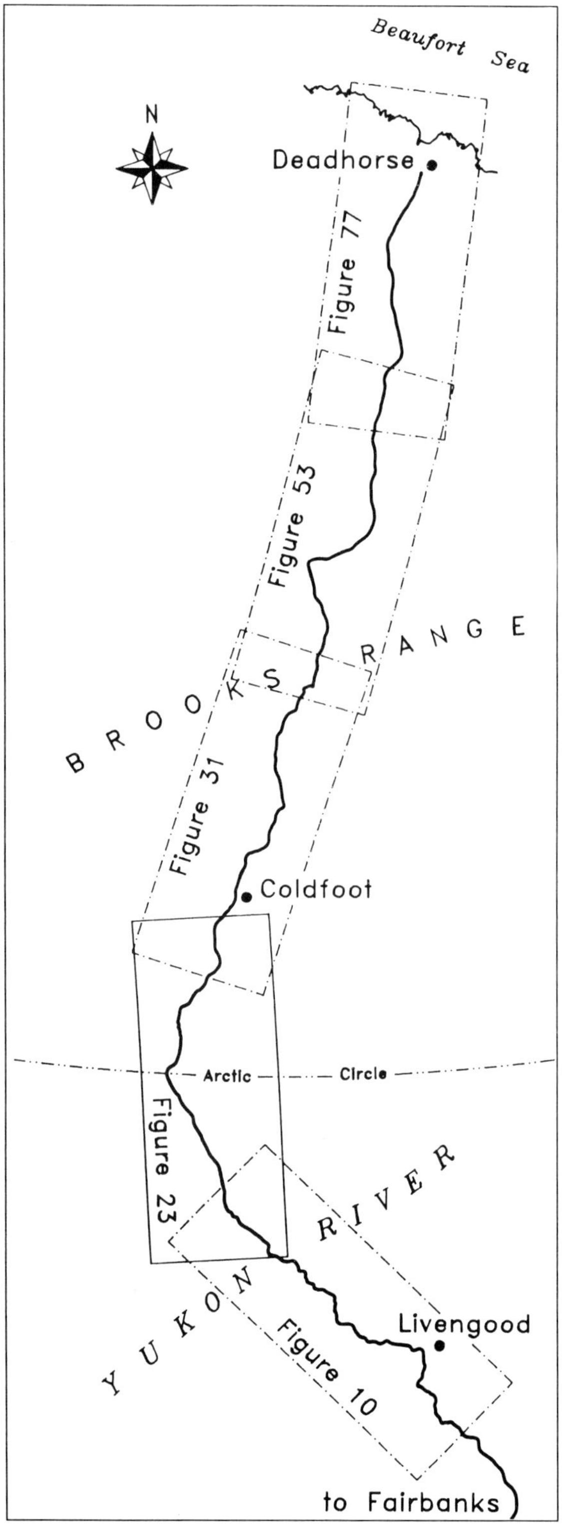

contains many small cavities.

These relatively recent lava flows may be related to flows on the north side of the Yukon Flats basin to the east. This basalt is overlain by several feet of tan loess.

Mile 79

No Name Creek. To the north, in the foreground, is a tussock-covered valley, an indication of underlying, ice-rich permafrost. A poorly-drained black spruce forest is in the distance.

Mile 82

Good view of erosional landforms called **cryoplanation** terraces to the north. Meltwater from overlying snow percolates into the soil layer, moving it downslope by freeze-thaw action and water washing. **Frost action** breaks up the exposed bedrock, which is then moved downslope through meltwater washing and **mass movement**. The local permafrost layer controls the horizontal expansion of the floor of the **terrace**.

Castle Mountain to the northeast is composed of schist and banded **quartzite** of the Ruby terrane.

Mile 86.3

Entrance to a quartzite quarry containing massive and iron-stained banded quartzite and schist of the Ruby terrane. Bedrock west of the quarry is **frost-shattered** and arranged by frost action into sorted **stone nets** and **polygons**.

Mile 87

Mackey Hill.

Mile 88.6

Possible open-system pingos in poorly-drained valley floor to the west.

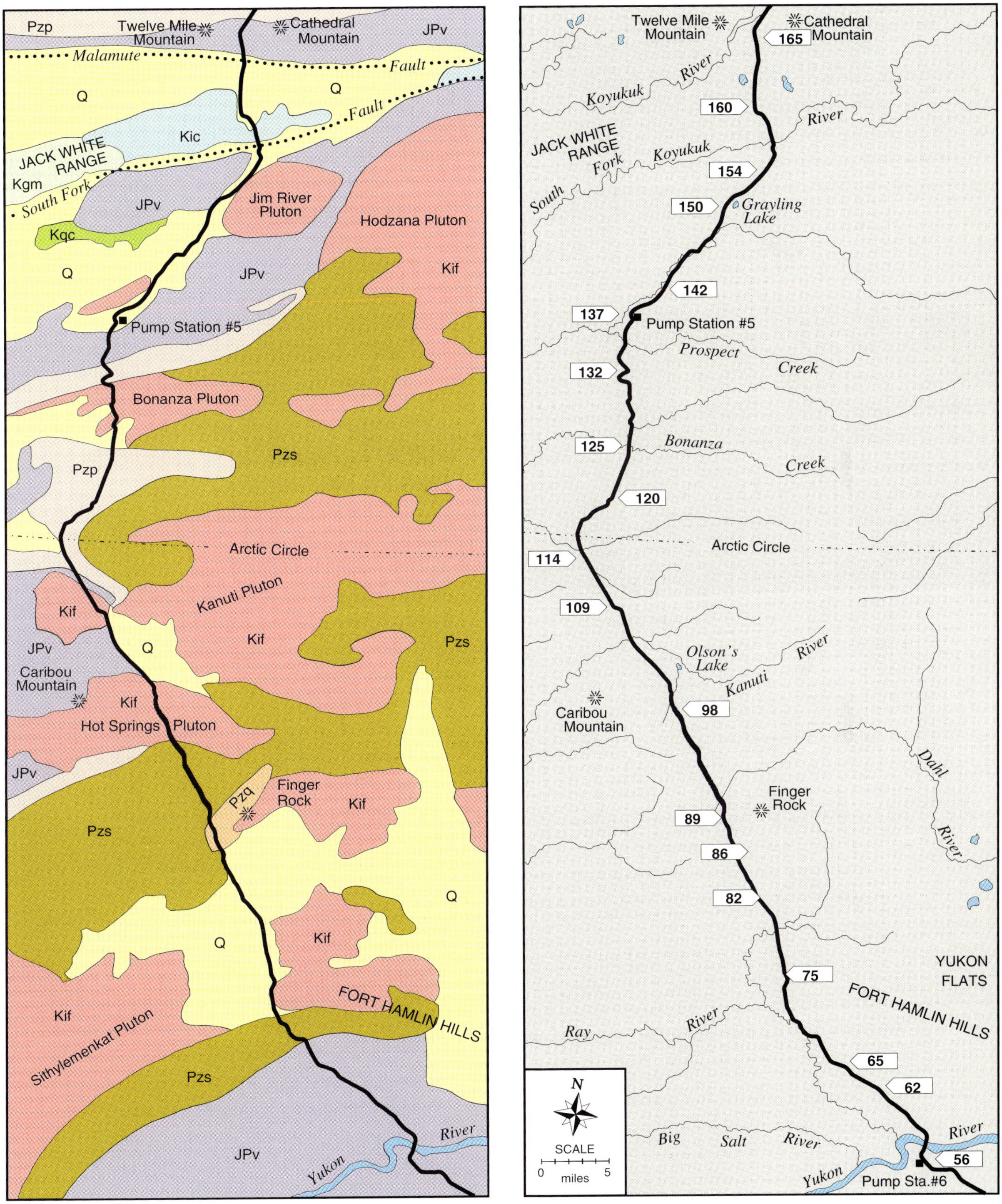

Figure 23. Dalton Highway geology and points of interest. Mile 80 to Mile 165

(from Brosgé and Reiser (1964) (1971), Brosgé and others (1973), Patton and others (1973) and Dillon and others (1986). See figure 92 for key.)

Mile 94.0

Boulders of schist and banded gneiss lie along an access road heading west toward a quarry. The metamorphic minerals in these rocks indicate that the these rocks as well as the surrounding country rock have been extensively altered by the intrusion of nearby granite **plutons**.

Figure 24. Granite tors on Finger Mountain.

Mile 98.5

Finger Mountain. The conspicuous rock pinnacles protruding from the nearby landscape are granitic tors of the 110 million-year-old (Cretaceous) Hot Springs pluton (figs. 24 and 25). Close examination reveals that the granite of these bedrock monoliths contains large **phenocrysts** of feldspar, a constituent mineral of granite. On the east side of the highway, Finger Rock, one of the more prominent granite tors, protrudes upward at an angle and resembles a pointing finger. During the pioneer days of Alaskan aviation, bush pilots used Finger Rock as a navigational aid.

The tors of Finger Mountain are surviving remnants of large, unglaciated bedrock hills subjected to intense **frost wedging** during periods of rigorous climatic conditions. These conditions are characterized by extensive areas of late snow, low summer temperatures, and frequent freeze-thaw cycles. Periglacial processes have been the main agents of landform modification in the area surrounding the tors. (The Scottish moors formed in a similar fashion.)

The joints (fractures) in the granitic rock that form most of the tors in the area provide easy access for moisture to penetrate the rock. As the water freezes, it expands and mechanically breaks the bedrock. The resulting debris moves downslope from its source by solifluction and other mass-wasting processes.

Solifluction is the slow, but effective, downslope movement or flow of water-logged

Edward Bovy

Figure 25. Finger Rock, one of the more prominent granite tors along the Dalton Highway, protrudes upward at an angle and resembles a pointing finger. The tors of Finger Mountain are surviving remnants of large, unglaciated bedrock hills subjected to intense frost wedging. During the pioneer days of Alaskan aviation, bush pilots used Finger Rock as a navigational aid.

sediments within the active layer above permafrost. Its formation is related to the slow evaporation rate within the active layer and the inability of excess water to penetrate the underlying permafrost. The front of a solifluction lobe often contains denser vegetation than the surrounding area and this helps make them easier to identify. Its most important effect may be the smoothing and rounding of hills.

Local sorting of debris and patterned ground develops where reduced slope angles restrict movement by solifluction. Sorted cobble and boulder polygons and **nets** lie adjacent to the road.

Mile 103.5.

Caribou Mountain to the west. The bedrock forming Caribou Mountain is composed

Figure 26. Old Man Camp during the pipeline construction days, approximately 1974.

of metamorphosed and partially serpentinized gabbro of the Angayucham terrane (Am). These rocks are thrust-faulted over the younger Cretaceous granites.

Test pits and drilling in the Kanuti River region indicate the potential for several million cubic yards of ore-grade **placer** tin (cassiterite) deposits. Cassiterite is one of the few tin-bearing minerals in the world and is the principal ore of the metal. It typically occurs in veins located within or close to granite masses.

Mile 105.9

Old Man Camp (fig. 26).

Mile 109

Road cut exposes granite of the Kanuti Pluton. An accumulation of fragmented rock or **grus** overlies the bedrock and is produced by the disintegration or crumbling of the granite. The top of this hill marks the contact between the mafic **igneous** bedrock of the Angayucham terrane to the northeast and the granitic bedrock to the southwest. Numerous outcrops of weathered porphyritic granite lie on the hilltops west of the road.

Mile 111

Well-developed **solifluction lobes** to the southwest.

Mile 113.9

The Dalton Highway intersects the Hickel Highway, a winter road constructed in 1969 to haul supplies north to the Arctic Slope. The ice road was built by compacting hard snow and smoothing uneven road surfaces. This quick and simplistic approach to road construction in a permafrost environment proved costly. By the end of the first summer following construction, the roadbed had rapidly deteriorated because the insulating surface-vegetation mat had been destroyed by the heavy equipment used to build the road.

USDI BLM photo / Edward Bovy

Figure 27. BLM's interpretive wayside exhibit at the Arctic Circle also serves as a rest stop complete with picnic tables and outhouse.

Mile 114

Fish Creek - Placer gold may be found within one mile of the bridge on either side.

Many streams along the Dalton Highway are open to recreational mineral collection by methods not requiring a permit. Any area in the **Utility Corridor** outside the pipeline right-of-way (27 feet either side of the pipeline) that is not staked and legally claimed for mining or other purposes is open to recreational mineral collection. You may use metal detectors, gold pans, or manually fed **rocker** or **sluice boxes** to recover the gold.

Mile 115

The Arctic Circle (fig. 27) is an imaginary line, drawn parallel to the equator at latitude 66°33'32" N. At this latitude, seasonal tilting of the earth's northern hemisphere away toward the sun (summer months) produces at least one 24-hour period of daylight (including twilight hours). The total consecutive 24-hour periods of daylight increase northward up to a maximum of 180 days at the North Pole. The reverse holds true for winter months when at least one 24-hour period of darkness persists when the northern hemisphere tilts away from the sun.

The bedrock in this area consists of phyllite and **slate** of the Ruby terrane.

Mile 119

Solifluction lobes on the southeast side of the road. Tors of the Kanuti pluton are visible on the ridge top to the east.

Mile 120

The Koyukuk River basin, a lowland to the west, was inundated by glacial ice probably during late Tertiary or early Quaternary time. Two of three glacial periods recognized within the Brooks Range are represented in this area: the Gunsight Mountain **erratics** and the younger Anaktuvuk River Glaciation (fig. 28).

Geology and the Arctic Circle...

Technically speaking the Arctic Circle actually changes over time. As the earth rotates on its axis, it wobbles back and forth like a spinning top due to the gravitational torques of the sun, the moon and the planets. This cyclic phenomena repeats about every 25,800 years; this means that one year midway through the cycle, every place on earth will have 12 hours of daylight every day for an entire year, and the North Star will switch from being Polaris to Sirius!

Another closely-related phenomena is polar wandering. The magnetic pole (which attracts our compass needles) does not coincide with the geographic North Pole. Right now, the magnetic pole is located at 77°20' N, 101°48"W, on Bathurst Island in the Canadian Arctic Islands Archipelago. Measurements during the last 150 years show that it has wandered there from 70°N, 97°W (the north-central part of Canada's Northwest Territories). This polar wandering is due to fluctuations and changes of the earth's magnetic field and is probably caused by movements within the earth's core, and to a minor extent, the drift of the continents.

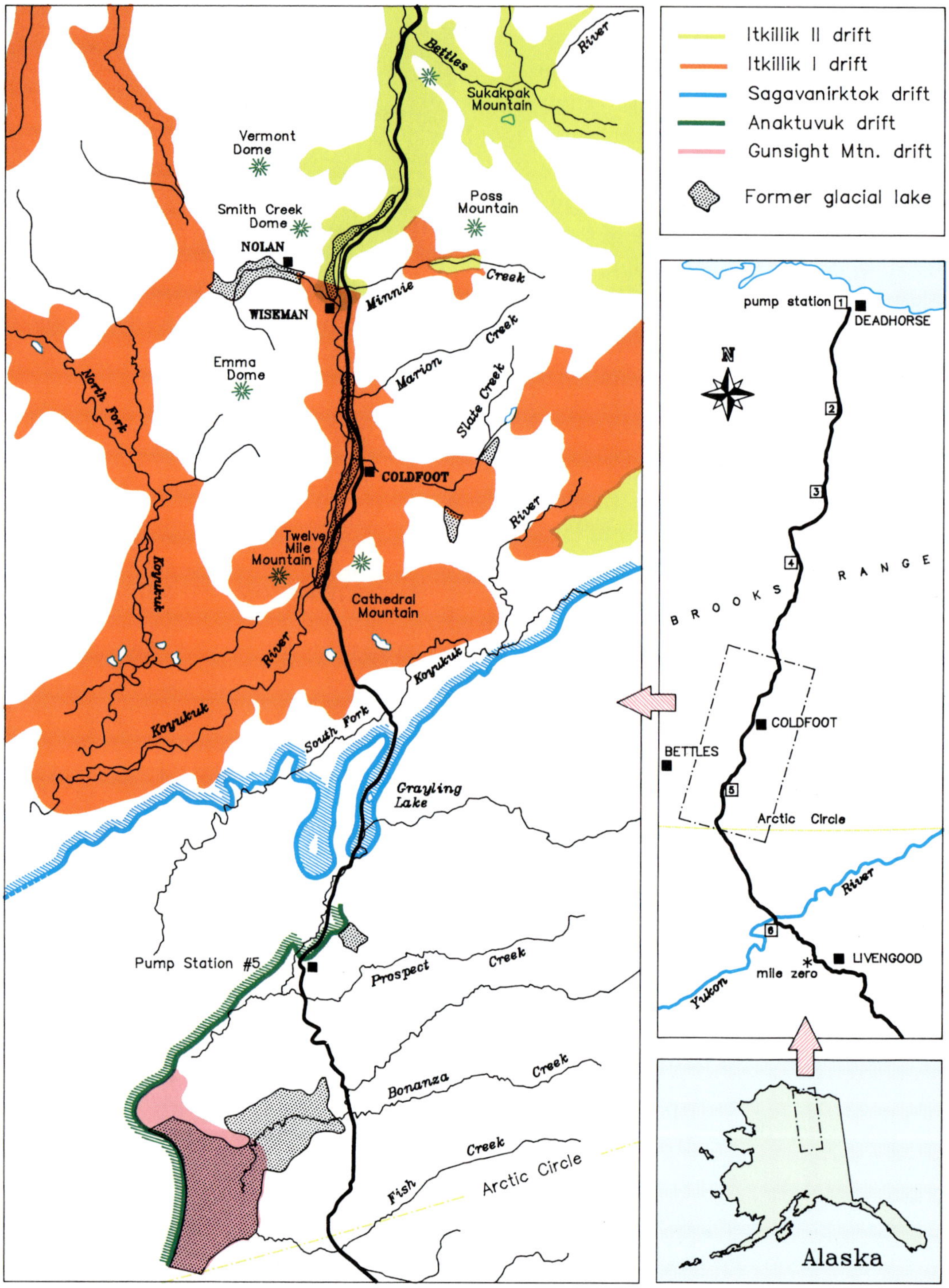

Figure 28. Glacial limits of the area between the Arctic Circle and Sukakpak Mountain.
(after Hamilton, 1978b, 1979b and 1982)

The Gunsight Mountain erratics represent the oldest of three principal glacial intervals in the central Brooks Range. This interval is defined by weather-resistant erratic boulders that extend as much as 15 miles beyond the Pleistocene drift limits.

The Anaktuvuk River Glaciation, the second major glacial interval, was preceded and followed by major interglacial periods of uplift and erosion. Drift from this interval has been greatly modified by stream erosion and by periglacial processes.

The Anaktuvuk River Glaciation produced a **proglacial lake** that filled the valley of Bonanza Creek to about the 800-foot contour. The glacier-dammed lake was fed by meltwater streams flowing out of the glacier front and by the drainage waters of Fish and Bonanza Creeks.

Mile 125

South Fork Bonanza Creek. Placer gold may be found on either side of the bridge outside the pipeline right-of-way. Placer claims begin 1.5 miles upstream from the bridge.

S.F. Bonanza Creek marks the approximate contact between the metamorphic rocks of the Ruby terrane and the Cretaceous granite of the Bonanza pluton.

Mile 130-132

Quartz-mica schist bedrock is exposed along the road leading to the Gobblers Knob turnout.

Mile 132

Gobblers Knob turnout (with outhouse). View of the South Fork Koyukuk River basin, the Jack White Range to the northwest, and Pump Station 5 to the north.

Mile 134

Prospect Creek. Placer gold may be found

YUKON-KOYUKUK BASIN

The Yukon-Koyukuk basin is a triangular lowland that extends from central Alaska to the Yukon delta and Norton Sound. At its base, the basin includes sediments eroded from volcanic rocks. Quartz-rich, graywacke sandstones, representing deposition from sediment-laden currents in a deep-water basin, overlie and interfinger with the volcanic sediments.

Non-marine sedimentary rocks, some of them having coal (such as at Tramway Bar on the Koyukuk River, 3 miles west of MP 160), overlie the other sediments and crop out around the edges of the basin. These sedimentary rocks are well-indurated (that is, they hold together well) and are often interlaced with quartz veins.

The basin has experienced near-metamorphic conditions through deep burial or from proximity to igneous intrusions, or both.

The Dalton Highway crosses the northeasternmost exposures of the Yukon-Koyukuk basin between mile 152 and mile 162. A quarry just north of the South Fork of the Koyukuk River provides the best exposures along the highway. This quarry exposes a dipping sequence of interbedded conglomerate, sandstone, and siltstone. These coarse-grained sedimentary rocks were deposited in high-energy environments at the margin of the basin.

The conglomerate contains mostly pebble-size to boulder-size rocks supported by finer-grained particles. Silicified graywacke sandstone, mafic volcanic rock, chert, and granite make up the major part of the fragments in these sedimentary rocks. The positions of these fragments through the sedimentary rock column record the progressive uplift of the area during the early and middle Cretaceous.

in areas outside the pipeline right-of-way both upstream and downstream from the bridge. Placer claims begin 1.5 miles upstream from the bridge.

Mile 135.7

Prospect Creek Camp. Placer gold is reported in an area extending from the end of the road at old Prospect Creek Camp to two miles downstream.

Mile 137.1

Pump Station 5 has a refrigerated foundation built on a ridge of cemented gravel flanked by ice-rich organic silt. (fig. 29)

Mile 142.3

Douglas Creek. Annual icing conditions along Douglas Creek flood this section of the highway during winter months. Icings, or "aufeis," are thick masses of surface ice formed in winter by the successive overflow and subsequent freezing of sheets of surface water or emerging ground water (fig. 30). As ice thickens in a stream channel, hydrostatic pressure increases beneath the ice and lifts it temporarily. Water eventually breaks through the ice cover, spreads out and freezes. This process, when repeated many times throughout the winter, may produce erosional problems and severe flooding by water as well as by ice. Special consideration was given to this phenomenon during early stages of pipeline and highway route planning and construction.

Mile 145 - 155

This area adjoins several geologic terranes. The hills are composed of Angayucham mafic rocks intruded by the Jim River pluton, located on the east side of the road. The small low hills along the west side of the highway are Angayucham mafics in fault contact with and

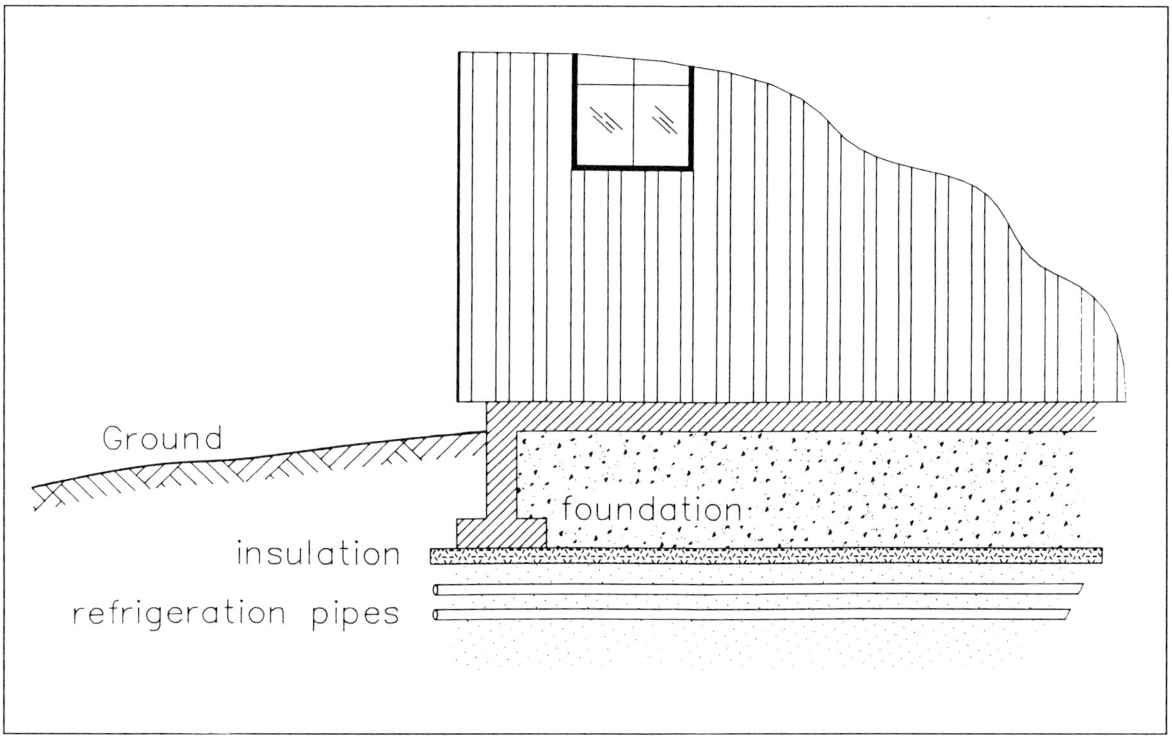

Figure 29. A refrigerated foundation at Pump Station 5 vents heat from the building to keep underlying permafrost frozen.

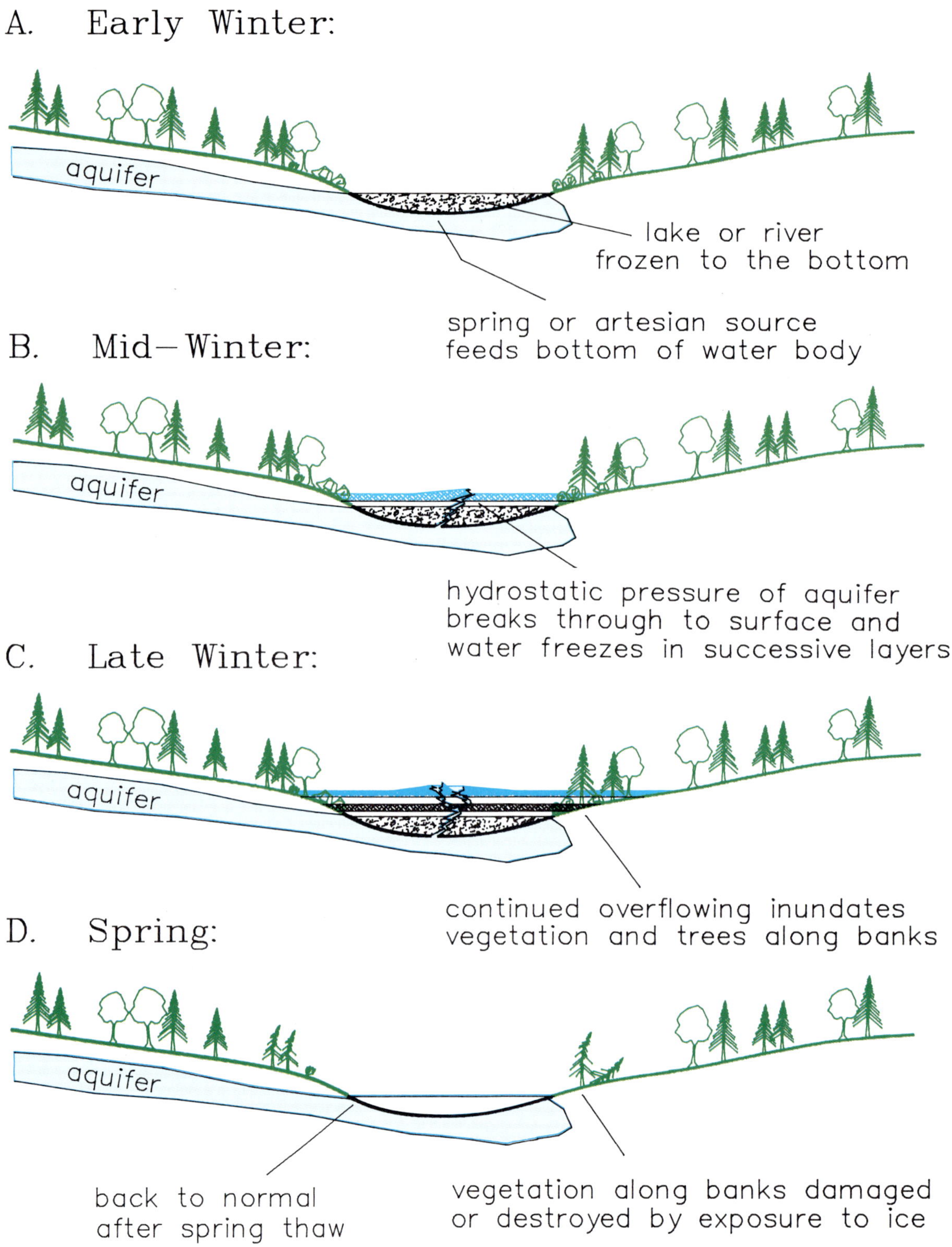

Figure 30. Aufeis (icings) development.

underlying the Cretaceous **clastics** of the Yukon-Koyukuk basin. Both units are in fault contact with the metamorphic rocks of the Ruby terrane.

Mile 145.2

The road climbs and crosses a probable **moraine** covered by black spruce trees. The drift is assigned to advances of the Sagavanirktok River Glaciation.

In the central Brooks Range the youngest interval of glaciation is represented by drift of the Sagavanirktok River Glaciation and by drifts I and II of the Itkillik Glaciation. This interval was not interrupted or followed by major erosional periods and because of this, drifts from this interval have retained much of their original topographic form.

Mile 145.5

Road passes over elevated pipeline.

Mile 150.5

Grayling Lake is a shallow body of water that probably formed as a thermokarst lake in older lake sediments that fill the valley at this location. The lake lies in a **U-shaped**, glacially-scoured trough of Sagavanirktok River age that has been greatly modified by postglacial **weathering** and erosion. Solifluction lobes, **talus cones**, and **alluvial fans** are located along a slope formed by the Jim River pluton to the east.

Mile 154.2

Jim River Pluton to the east consists of **monzonite**, **syenite**, and a core of granite. Boulders and rubble of both rock types are found in a small quarry and creek bed east of the highway. This pluton is one of the many Cretaceous plutons that intruded the Ruby geanticline about 105 million years ago.

Intense seasonal **frost splitting** during

Refer to figure 31 for geology and points of interest between Mile 165 and Mile 255.

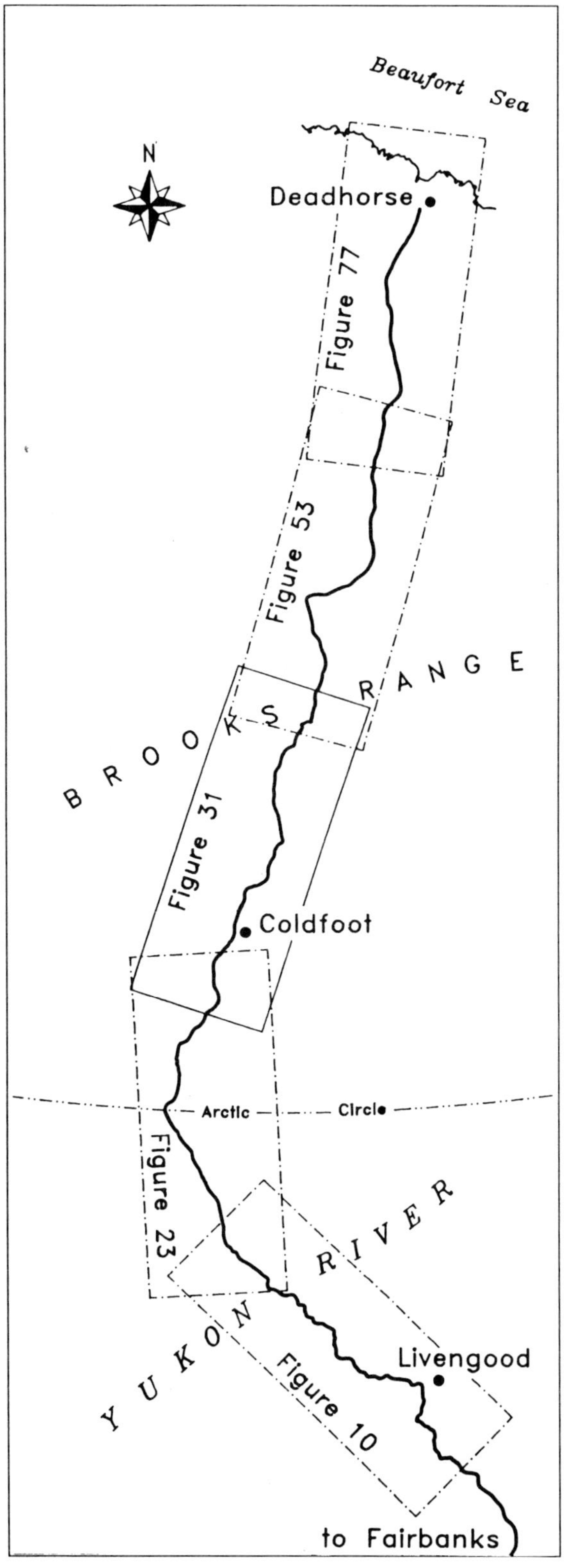

the last glacial period produced the coarse bedrock rubble (referred to as a rubble sheet) along the west-facing slope.

Mile 156.3

South Fork of the Koyukuk River. The South Fork fault, which trends east-northeast along the river valley, marks the northern limit of the Cretaceous granites in Alaska. This fault locally separates the Angayucham terrane from the Ruby terrane. It is one of many Cretaceous, east-west, high-angle (greater than 45 degrees) fault zones exhibiting many miles of horizontal movement.

Placer gold may be found 1.5 miles upstream and two miles downstream in **bench gravels** located on the north side.

Mile 157.7

Road east leads to a rock quarry composed of hardened sandstone, siltstone and prominent pebble-to-cobble conglomerate. The conglomerates, which are poorly sorted, are supported in a matrix of mud and sand, indicating rapid and chaotic deposition. Locally, channels cross cut sandstone and siltstone beds.

The individual fragments, or clasts, that form the sedimentary rocks at the bottom of the section are mostly rounded, mafic igneous rocks. Clasts near the top of the section are formed more from schist. The change in clasts, from mostly igneous to metamorphic, probably reflects the sequential uplift of terranes from which they eroded.

Mile 159

Pipeline crosses under road.

Mile 159.5

The highway intersects **terminal moraines** of the Itkillik I Glaciation, the youngest of several major glacial periods in the Brooks Range.

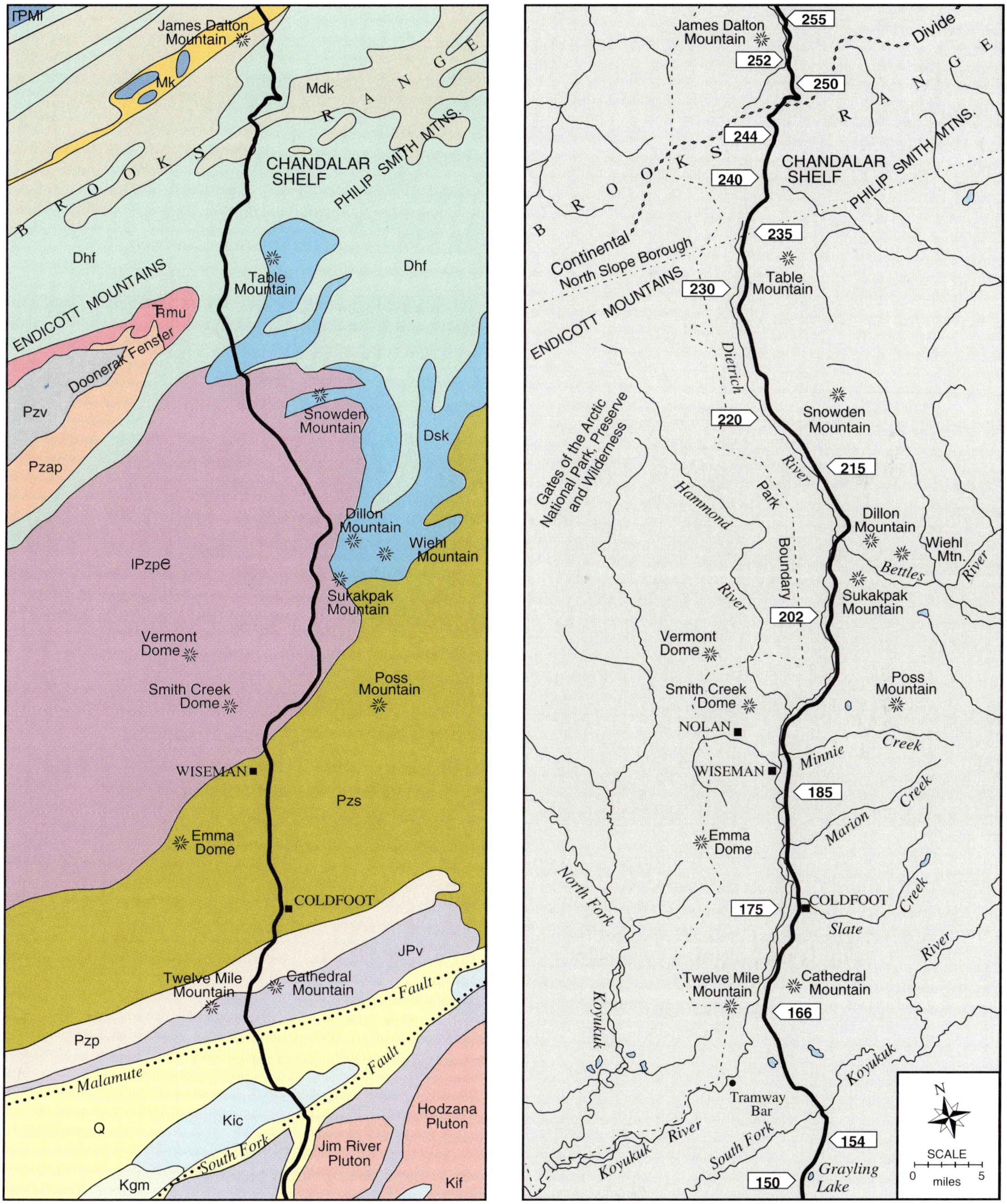

Figure 31. Dalton Highway geology and points of interest. Mile 165 to 255

(from Brosgé and Reiser (1971), Chapman and others (1971), Brosgé and others (1979), Patton and others (1973), Dillon and others (1986), and Mull and others (1987). See figure 92 for key.)

Figure 32. Water-filled thermokarst pits form through melting of ice-rich ground in the upper permafrost layer. Caving of the ground layer progresses as support from the ice is removed through thawing.

Mile 164.5

Short section of buried pipeline.

Mile 164.7

The road crosses the approximate trace (not visible on the surface) of the east-west trending Malamute fault of mid-Cretaceous age. This fault along with the South Fork fault (MP 156.3) may be part of the larger Tintina fault system to the east. The Malamute fault has undergone about 25 miles of lateral movement.

Mile 166

The melting of ice-rich ground in the upper permafrost layer produced the water-filled thermokarst pits on the west side of the road (fig. 32). Caving of the ground layer progresses as support from the ice is removed through thawing. Thermokarst development is commonly initiated by local changes leading to warmer ground temperatures.

Mile 168

Cathedral Mountain to the northeast and Twelvemile Mountain to the west are composed of Devonian-to-lower-Jurassic volcanic rocks and Mississippian-to-lower-Jurassic chert of the Angayucham terrane.

Mile 166.5

Pipeline crosses under road.

Mile 170

Rosie Creek. Excavation of a near vertical road cut on the south side of Rosie Creek in 1975 exposed **ice wedges** in fine-grained solifluction deposits derived from glacial **till**. Ideally, the organic mat extends farther downslope as the ice-rich soils thaw and the slope continues to recede. Its insulating effect retards further melting until the slope reaches a relatively stable state.

Small hummocks, or cemetery mounds, produced by melting ice-rich sediment and ice wedges, have formed on the west side of the road, opposite the roadcut.

Two small creeks one mile south of Rosie Creek on the east side of the highway contain placer gold. A road to the east leads to a rock quarry of gabbro that contains **pyrite** and **chalcopyrite** minerals.

Mile 170

Solifluction lobes to the east.

Mile 171.7

The road parallels a moraine formed during the Itkillik II glacial advance which terminated in a lake that formed behind the southernmost Itkillik end moraine belt.

Mile 172.5

View looking north toward the town of Coldfoot. A road to the east leads to a rock quarry containing large boulders of mafic igneous rocks composed of black, shiny crystalline gabbro and dark gray, mostly fine-grained **amphibolite**. Look for cubes of pyrite and chalcopyrite.

Mile 175

Coldfoot. During the late nineteenth century, the ever-persistent prospectors of Alaska discovered gold placer deposits on several tributaries of the Middle Fork of the Koyukuk River. The town of Coldfoot, originally named Slate Creek, was the site of one such discovery about 1899. According to Marshall (1933), one of the waves of early gold prospectors who traveled up the Koyukuk to this point got "cold feet" upon winter's arrival and soon departed. By 1905, 130 men were working 28 mines in the area and the year's mining production totaled $200,000.

The boom-to-bust cycle, an often repeated

USDI BLM photo / Edward Bovy

Figure 33. Early 20th century cabin at Coldfoot, Alaska.

event throughout Alaska's history, struck a heavy blow to the Coldfoot economy when mining interests shifted to new gold discoveries 15 miles up river on Nolan Creek. By 1907, the once bustling mining camp of Coldfoot was all but abandoned (fig. 33).

Almost 60 years later, prospectors of a different sort tapped into huge oil and gas reserves at Prudhoe Bay on the edge of Alaska's Arctic coast. Again the promise of riches from the earth, this time in the form of "black gold", brought new life to the community of Coldfoot. This new "gold rush" came in the form of jobs related to construction of the Dalton Highway and the Trans-Alaska Pipeline System which began in 1974. The boom continues today as freight traffic and seasonal visitors pass through historic Coldfoot each day.

Gold placer mining is still a way of life for many Coldfoot area residents. There are over 2,000 active mining claims (as of 1991) within the Wiseman and Chandalar quadrangles alone.

Mile 175

Slate Creek. The gold-bearing bench and shallow creek gravels of Slate Creek were discovered in 1899. Myrtle Creek, an upstream tributary of Slate Creek, became one of the biggest placer-gold producers of the Koyukuk district. Although total gold production is unknown, about 8,800 **troy ounces** of gold were recovered between 1900 and 1910. Some of the gold was very coarse. For example, a nugget weighing nearly 23 troy ounces was unearthed in 1940.

Placer gold may be found 0.25 to 0.75 miles upstream from the bridge.

Mile 176

Clara Creek. Some placer gold mining reportedly took place here around 1901 and again in 1934. Placer gold is found 0.5 miles

ARCTIC ALASKA TERRANES

The Arctic Alaska terranes (AAC, AAE, AAH, AAN; see fig. 4) are a complex assemblage of at least four distinct belts, provinces or subterranes. These terranes comprise the central portion of the Brooks Range. The mountains demonstrate the effects of the compressional forces which uplifted and amalgamated or "welded" these rocks. However, not all the stratigraphic and tectonic relationships between these provinces are above speculation. In two of the provinces, some of the internal stratigraphy isn't even agreed upon. From south to north you cross the Coldfoot Schist Belt, metamorphosed lower Paleozoic rocks, Devonian sedimentary rocks, and the North Slope succession.

Unlike the previously described terranes, Arctic Alaska terranes are composed of mostly thick, distinct, and interbedded sedimentary or metamorphosed rocks. Some of these are typically very hard and commonly weather to ridge-forming outcrops. Such characteristics make many of these units readily identifiable as named formations along the Dalton Highway.

Also along the road you can see outcrops showing thrust-fault mechanics at work. Fault imbrications, repetitively superimpose multiple stratigraphic sections of the thick and massive bedded, structurally competent, strata upon each other, rather like books leaning against each other. Thus, rocks from different formations or terranes may be in contact across a fault. Sometimes they may overlie one another. Also evident is how interbedded and less competent strata react to compressional **tectonics** by acting as slide planes (or décollements) for thrust faults. These units also show the development of intraformational folds. Some of these displays are quite spectacular!

Coldfoot Schist Belt

The Coldfoot Schist Belt is one of the areally smaller terranes encountered along the Highway. It is elongate to the northeast and intersects the road between mile 172 and 187 (fig. 4). These are mostly silvery-to-black graphitic schists with quartzite lenses or layers. There are minor amounts of **marble**, volcanic rocks, mafic igneous rocks and some local igneous intrusions which have altered the sedimentary carbonate rocks to **skarn**. The rock fabric and mineralogy tells us that these have undergone multiple phases of intense metamorphism.

Placer Deposits

One of the most valuable placer minerals is gold, the heavy and brightly-colored metallic element that has lured tens of thousands of prospectors to Alaska during the last hundred years.

Placer deposits begin their formation through a destructive process called weathering. Weathering of the gold-bearing bedrock takes place in the form of physical disintegration and chemical decomposition. Once released from the host rock, the heavier placer materials begin to concentrate as the surrounding lighter and finer materials are removed through erosion. At this point, little or no transportation of the placer material takes place. However, as conditions change, these deposits may be moved to a new and sometimes distant location by such transporting agents as water or ice.

There are many types of placer deposits, but one of the most important to the Alaska prospector is the stream placer. The action of flowing water effectively concentrates the heavier placer material while removing the lighter, worthless materials downstream. Gold particles carried downstream are deposited on the inside curves of a stream where water velocity has been reduced. It may also be trapped in the cracks and crevices of bedrock lying beneath the alluvium. Bedrock traps harbor the highest concentration of placer gold particles simply because the gold cannot migrate further downward.

Placer Mining

Placer mining is a surface mining technique for recovering placer minerals such as gold. This technique is used along abandoned stream gravels, in stream valleys, on beaches and underwater offshore marine environments, and in rock-outcrop equivalents of these environments. Once the placer deposit is identified, the overlying, non-gold bearing material, or overburden, is removed and the gold-bearing material is transported to a wash plant where water sorts the material by size. The smaller sized material is then mixed with water flowing through a sluice box. The gold sinks to the bottom of the box while the lighter material, still in suspension, is carried away by the flowing water.

Large-scale placer-mine operations use heavy equipment such as bulldozers and front-end loaders to handle the material. Small-scale operations use smaller bulldozers and loaders or suction dredges to move and process the gold-bearing gravels. The ever-popular gold pan is still used, but is a better sampling tool than a mining technique since an experienced panner may work only about a cubic yard of gravel per day.

Drift Mining of Placer Deposits

Drift mining is an underground mining technique that is used to extract frozen or stable gold-bearing gravels. The non gold-bearing gravel, or overburden, is left undisturbed and only the gold-bearing gravel in the **pay streak** and a small amount of the bedrock is mined. Equipment ranging from picks, shovels and wheelbarrows to mechanized apparatus are used to excavate the gravel and haul it to the processing plant. This method of mining placer gold has been used quite successfully in Alaska's interior region since the turn of the century and is particularly well-suited to areas that contain permafrost and deeply buried gravels.

In a typical Alaskan drift mine operation, a vertical shaft is sunk in frozen gravel until bedrock is reached. At this juncture a horizontal passage, or **drift**, is excavated in the gold-bearing gravels, just above the bedrock surface. A portion of the upper bedrock surface is also mined because it too may have gold particles lodged in cracks and **crevasses**. The perennially frozen gravels provide internal support for the walls and roof of drift and care must be used not to allow thawing along these surfaces. For this reason, drift mining usually takes place during the winter months. The excavated pay gravels are transported to the surface where they are processed during the spring and summer. Reclamation of the small amounts of tailings is completed during late summer and fall, after all of the gold-bearing gravels have been processed.

upstream on the east side of the highway.

This portion of the highway was washed out in 1989. The roadbed was subsequently raised an additional four feet to prevent further washouts.

Travel tip: the 28-site Marion Creek Campground was constructed by the Bureau of Land Management in 1993. It is the first developed public campground along the Dalton Highway and has sites for tenting and recreational vehicles. Fees are charged.

Mile 180

Marion Creek. Placer gold has been reported in the surface gravels. About 50 troy ounces of gold were recovered from the gravels of Marion Creek between 1901-1909.

Mile 181-183

Emma Dome on the west side of the road represents the core of an **antiform** that trends east-northeast across the Dalton Highway near Marion Creek. The antiform is composed of Paleozoic or **Proterozoic** schist intruded by gabbro and rhyolite-dacite dikes.

Mile 184

Dry Gulch. Placer gold may be found upstream on the east side of the highway.

Mile 185

Massive rock slide on Kalhabuk Mountain to the west.

Mile 185.5

View looking toward Wiseman, a small mining community founded in 1907 after gold was discovered on a narrow strip of land near Nolan Creek. The discovery prompted most Coldfoot prospectors to relocate the additional 15 miles up the Koyukuk River. This exodus left Coldfoot virtually abandoned. Many buildings from Coldfoot, including cabins and the schoolhouse, were eventually dismantled and rebuilt in Wiseman.

Mile 185.9

Access road to quarry on east side of road. Here, mica schists are folded from different episodes of metamorphism.

Mildly Metamorphosed Lower Paleozoic Sediments

From mile 187 to about mile 224 (on the west side), the Dalton Highway traverses mostly lower Paleozoic rocks. The regional rock fabric suggests that several phases of metamorphism have affected this succession of rocks. However, much of the bedding and rock fabric has been preserved, which means that the metamorphism was less intense than that in the Coldfoot Schist Belt.

Along the Highway, are a succession of very hard, Cambrian or older, **calcareous** schist, quartzites, or marble with some Cambrian-Ordovician volcanic rocks and assorted phyllites. These are overlain by Devonian rocks, which some have mapped as the Whiteface Mountain Volcanics. They consist mostly of metamorphosed volcanic clastics or breccias with calcareous or chloritic schists. (Note thatthese are quite unlike rocks from the Angayucham or Tozitna terranes.)

The rounded hills seen along the highway are composed of the Whiteface Mountain metamorphic rocks. The most conspicuous unit in this area is the Skajit Limestone. The Skajit consists of up to 2000 feet of massively bedded, white limestone or marble which weathers to prominent cliffs, castelated ridges, and pinnacles. The Skajit is overlain by and interbedded with, or both, the Beaucoup Formation, a mostly black slate or graphitic phyllite with minor amounts of carbonates and conglomerates.

There is still some uncertainty as to stratigraphic relationships and depositional origins within this succession. Most evident along the Highway, the prominent Skajit Limestone and the Whiteface Mountain rocks thin dramatically to the west. Neither lithology is geographically extensive west of the Highway.

It is also not clear whether the Skajit interfingers with the lower Beaucoup west of the highway or whether thrust faulting juxtaposes these two units. From west to east, the units change from the older Cambrian sandstone and quartzite beneath the Beaucoup to schists underlying the Skajit Limestone. These changes between outcrop relationships suggest that the rocks were derived from a land source located to the east.

USDI BLN photo / William Diel

Figure 34. Cabin at Wiseman, Alaska.

Mile 186

The glacially scoured, U-shaped Koyukuk River valley, bound by 6,000-foot-high ridges of Paleozoic limestone, sandstone, shale, phyllite and schist, is typical of central and southern Brooks Range geology.

The road crosses an **end moraine** deposited by a late Itkillik glacier that flowed south out of the Brooks Range. The end moraine blocked glacial melt water drainage and formed a lake that has since drained. **Lacustrine** sediments from the lake are now overlain by 12 to 15 feet of **Holocene** alluvium in the Middle Fork Koyukuk River.

Mile 187

Minnie Creek. The road cuts **morainal** deposits of the late Itkillik-age glaciation. Placer gold may be found on the east side of the highway about 1.5 miles upstream on Minnie Creek and along Larson Creek to its head.

Road crosses the trace (not visible on the surface) of the Wiseman thrust fault at Minnie Creek. This marks the north end of the Coldfoot schist belt.

Mile 188.5

Turnoff leads to the town of Wiseman (fig. 34). From 1900 to 1970, more than 330,000 troy ounces of placer gold was produced in the upper Koyukuk and Chandalar mining districts. The richest deposits reportedly lie in a mineral zone that cuts across the Dalton Highway from the historic towns of Nolan and Wiseman to the mining camps along Gold and Linda creeks.

Mile 190-191

Middle Fork Koyukuk River. Spur dikes were built to protect the pipeline and highway during periods of high water. Placer gold may be found on Hammond River about one mile upstream from its confluence with the Middle Fork of the Koyukuk River.

Mile 196

Nugget Creek. Placer gold may be found upstream about one mile.

Mile 197

Gold Creek. Look for the old miner's cabin on the northwest bank, complete with a sod roof (fig. 35). Placer gold was discovered here in 1900 and was mined sporadically through 1992. Gold production from 1900 through 1909 was about 11,000 troy ounces. Placer gold may still be found about two miles upstream from the bridge.

Angular fragments of stibnite with quartz stringers in them have been found in the stream gravels.

USDI BLM photo / William Diel

Figure 35. An old miner's cabin, complete with sod roof, sits on the northwest bank of Gold Creek.

Mile 197.5

Linda Creek. View looking north at till/ drift deposits in the roadcut.

Mile 198

River containment dikes to the east.

Mile 202

View to the northeast of the Devonian Skajit Limestone of Sukakpak Mountain (elev. 4,459'). The massive marble rock of Sukakpak Mountain was deposited in an intertidal (the region between extreme high and low tides) marine environment about 380 million years ago. Since that time the limestone has been subjected to high temperatures and pressures, or both, during burial and has metamorphosed or "changed form" into marble.

Fossils are rare in the Skajit Limestone, and this fact supports the idea that it was deposited in an intertidal zone because shell-

Edward Bovy

Figure 36. The massive marble rock of Sukakpak Mountain was deposited in an intertidal marine environment about 380 million years ago. Since that time the limestone has been subjected to high temperatures and pressures, or both, during burial and has metamorphosed into marble.

bearing creatures are also rare in similar intertidal environments today. The Skajit Limestone is underlain by black, graphite-bearing, Ordovician phyllite.

An arduous, but rewarding, day-long hike ascends the south side of Sukakpak Mountain where the Skajit Limestone is in fault contact with Paleozoic schist. Veins of crystalline quartz and stibnite, an ore of the metal antimony, are present in this fault zone (fig. 37). Cinnabar, a mercury-bearing mineral, and tetrahedrite (a copper ore), are also reported.

The summit of Sukakpak provides stunning views of the glacially carved Dietrich, Bettles, and Middle Fork Koyukuk river valleys and flood plains, as well as the Gates of the Arctic National Park and Preserve (figs. 38).

Mile 203-204

Ice-cored mounds, some of which are topped by trees, rise above the wet lower slopes of Sukakpak Mountain (figs. 39 and 40). These small mounds probably form by the freezing of ground water as it moves downslope within the active layer (the section of ground above permafrost which thaws in the summer and freezes again in winter).

The Devonian Skajit Limestone of Sukakpak Mountain is folded into a **syncline** (a U-shaped fold in which the strata or limbs dip inward toward the center). This fold, however, is **recumbent**, or overturned (fig. 41), and both limbs of the syncline point northwest.

Alternate freezing and thawing pries marble rock fragments loose from the 2,000-foot-high, west-facing wall of the mountain. These fragments tumble to the base of the near-vertical cliffs to form aprons of rock rubble.

Mile 204-209

The mountains visible along the west side of the road consist of thick gray Cambrian marble and mica schist, with lesser amounts of phyllite and dolomite.

Mile 207

Dillon Mountain, east of the road, is composed of south-dipping Skajit Limestone (fig. 42). The Skajit is folded into an overturned syncline similar to the fold at Sukukpak Mountain. Dillon Mountain is named in memory of John T. Dillon, an Alaska state geologist who spent most of his Alaska career studying Brooks Range geology.

Mile 208

Dietrich River crossing. The Dietrich River flows south 35 miles from its headwaters in the Endicott Mountains. It joins the Bettles

USDI BLM photo / William Diel

Figure 37. Tightly folded beds of stibnite-bearing marble of the Skajit Limestone crop out on the south side of Sukakpak Mountain. The marble is in fault contact with the underlying mica-quartz schist of Cambrian age.

USDI BLM photo / Arthur Banet, Jr.

Figure 38. View looking north at the glacially-carved Dietrich River valley. Note the gentle U-shape of the valley and remnants of aufeis along the Bettles River in the foreground.

Figure 39. Ice-cored mounds protrude above the wet, lower slopes of Sukakpak Mountain. These small mounds probably from by freezing of the ground water as it moves downslope within the active layer of permafrost.

Figure 40. Ice lenses or discontinuous layers of horizontal ground ice form in the lower slopes of Sukakpak Mountain.

River to form the Middle Fork of the Koyukuk River.

Mile 210.5

The hills on both sides of the Dietrich River consist of sandstone, siltstone, phyllite, and limestone of the Devonian Beaucoup Formation, overlying black Ordovician phyllite, rather than the Skajit Limestone as at Sukukpak Mountain. It is not clear whether this is due to erosion of the Skajit, the interfingering of differing rock units, or thrust faulting. The mountains on the west are composed of the Beaucoup Formation, and the mountains east of the road are mostly Cambrian metamorphic rocks of Ordovician clastic and carbonate lithologies.

Mile 212

A 270-foot-long section of pipeline buried near an undetected ice lens in an otherwise thaw-stable flood plain of the Dietrich River, had settled 15 feet vertically by 1985. A one-mile section of buried pipe was subsequently raised above ground to prevent further thaw-

ing and subsequent settling of the unstable soils.

Mile 215

The thermokarst lake on the west side of the road shows typical caved banks with trees and vegetation mat dipping inward toward the water as gradual melting of the surrounding ice-rich permafrost continues (fig. 43). Thaw ponds and lakes are thermokarst features formed through the decay of ground ice.

Mile 218.5

View to the east of Snowden Mountain (elev. 6,240 ft.), a marble **massif** formed from a faulted syncline composed of Skajit Limestone at its core. Fossils of **brachiopod** and trilobite (an extinct, possibly carnivorous, marine invertebrate) have been recovered at the northern base of Snowden Mountain. The Cambrian fossils are embedded in a brown weathered marble unit that is overlain by the Devonian Skajit Limestone.

The prominent spires on the north side of Snowden Mountain are Cambrian to Ordovician, calcium carbonate-bearing schists, phyllites and marble that have been thrust faulted northward onto the rocks of the Whiteface Mountain volcanics.

Mile 219

The unvegetated and braided flood plain of the Dietrich River to the west is the site of seasonal icing sheets up to 16 feet thick.

Mile 220.4

The conspicuous hill on the slope across the Dietrich River is an open-system pingo formed in till of late Itkillik age.

Mile 221.7

Skajit Limestone quarry with **calcite** and pyrite-bearing marble. This exposure of thick,

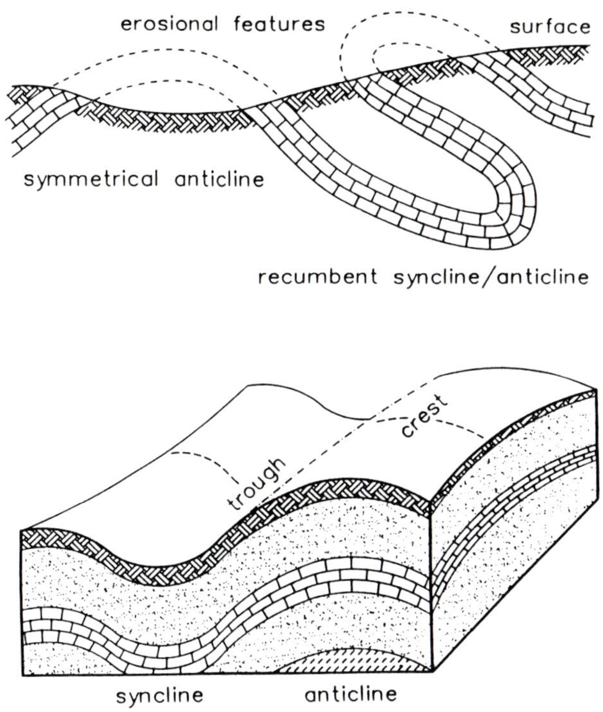

Figure 41. Block diagrams illustrating the principal types of folds
(after Judson and others, 1976)

USDI BLM photo / William Diel

Figure 42. View looking south at the rugged Skajit Limestone of Dillon Mountain.

Figure 43. A thermokarst lake near Milepost 215 shows typical caved banks. The trees and vegetation mat dip inward toward the water as gradual melting of the ice-rich permafrost continues. Thaw ponds and lakes are thermokarst features formed through the decay of ground ice.

dark gray, Skajit Limestone has been metamorphosed to coarse marble with minor amounts of pyrite. The quarry has thick calcite veins along **bedding planes** and numerous crosscutting calcite veins throughout.

On Table Mountain, to the northeast, the Skajit is 2,000 to 3,000 feet thick. The prominent cliffs of Skajit Limestone west of the highway are only about 100 feet high, and are interbedded with the Whiteface Mountain volcanic rocks.

Mile 224

Hillsides are composed of the southernmost exposures of the Hunt Fork Shale. These are brown and interbedded rocks which typically show folded and dipping beds. The Hunt Fork is exposed in an overturned **anticline** west of the highway.

East of the road, the Skajit Limestone and metamorphosed volcanic rocks are thrust over the Hunt Fork Shale.

Mile 225.8-226

In the western and central Brooks Range the Hunt Fork Shale and Kanayut Conglomerate typically form the base of a structural or tectonic **allochthon**, which may be thought of as a rather large sheet of rocks moved laterally great distances by thrust faults.

In the Doonerak Fenster area to the west, one of these allochthons is faulted over the top of a succession of the younger, middle Ellesmerian rock sequence of the upper Mississippian through the Triassic. An erosional window, or hole, called a fenster, exposes the underlying middle Ellesmerian sequence. Middle Ellesmerian rocks are typically found in the subsurface of the North Slope or along the northernmost exposures of the Brooks Range (figs. 4 and 5).

Looking west up Kuyuktuvuk Creek, there is a good view of the eastern end of the Doonerak Fenster.

The confluence of Dietrich River and Kuyuktuvuk Creek. Bedrock on the west side of the Dietrich River to about mile 235 is Hunt Fork Shale. A day hike up Kuyuktuvuk Creek takes you on a journey through a sequence of North Slope stratigraphy. This sequence of Mississippian Kekiktuk Conglomerate, Kayak Shale, and Lisburne Limestone **unconformably** overlies lower Paleozoic argillite. A thrust-faulted succession of Devonian Hunt Fork Shale and Kanayut Conglomerate overlies the North Slope succession. (Both successions are described in more detail later.)

Mile 227.2

Nutirwik Creek. River training structures on the east side of the road protect the buried pipeline during periods of high water. There is a good view of Skajit Limestone on Table Mountain to the northeast (fig. 44). A day hike up Nutirwik Creek (when water in the creek is low) will take you through a section of Skajit on the way to Chandalar Shelf. In a steep section near the pass, large complexes of metamorphosed volcanic rocks underlie the Skajit.

Mile 228.5

Roadcut in Hunt Fork Shale, east side of road. The upper Koyukuk-Dietrich River system has a steeper gradient and is more extensively braided than its middle reaches.

Mile 230

Debris flow on the east side of an unnamed stream is about 100 yards wide consists of silt and mud intermixed with cobbles and boulders. The flow brought down several trees and covered the roadbed in the spring of 1992.

Mile 235.2

The northernmost spruce tree on the Dalton Highway.

USDI BLM photo / William Diel

Figure 44. View looking northeast up Nutirwik Creek at Table Mountain. Alternate freezing and thawing pries marble rock fragments loose from the 1,000-foot high, west-facing wall of the mountain. These fragments tumble to the base of the near-verticle cliffs to form aprons of rock rubble. Note the wide flood plain of Nutirwik Creek and the relatively narrow channel it occupies during periods of low water.

USDI BLM photo / Arthur Banet, Jr.

Figure 45. View to the north of Chandalar Shelf, a southeast-trending, glaciated, western, tributary valley to the North Fork Chandalar River. Across the broad, drift-covered floor of the shelf lies a large southwest-trending ridge composed of late Pleistocene morainal material. The moraine was deposited by a glacier tongue that poured over Chandalar Shelf and down into the Dietrich River valley to the south.

Figure 46. A section of 48-inch pipe being lowered into position.

Mile 236.5

The highway climbs to the top of the Chandalar Shelf. A road-cut here exposes black, shiny phyllite of the Devonian Beaucoup Formation. Note that cross-cutting quartz veins are common and that some of the original sedimentary bedding is still visible in this mildly metamorphosed rock.

Turf-banked solifluction lobes are visible on the slope to the west. View of the Upper Dietrich River valley to the south and west.

Mile 237.5

Chandalar Shelf is a southeast-trending, glaciated, western, tributary valley to the North Fork Chandalar River (fig. 45). Across the broad, drift-covered floor of the shelf lies a large southwest-trending ridge composed of late Pleistocene morainal material. The moraine was deposited by a glacier tongue that poured over Chandalar Shelf and down into the Dietrich River valley to the south.

Alyeska Pipeline Service Co.

Hunt Fork Shale and Kanayut Conglomerate

The core of the Brooks Range is composed of sediments that were eroded from a now departed, northern landmass situated where the Beaufort Sea now exists. These rocks are called the Ellesmerian depositional sequence. The sequence records three distinct phases of sediment deposition from a northern source which took place from the Devonian through the Triassic or into the Jurassic. The Hunt Fork Shale and Kanayut Conglomerate form the basal portion of the Ellesmerian sequence.

The Hunt Fork and Kanayut are upper Devonian to possible lower Mississippian age. Together, they represent one of the most extensive and thickest **fluvial** sequences in North America. The nonmarine sandstone and conglomerate facies crop out along over 500 miles of the Brooks Range. These rocks record a complete depositional cycle. The Hunt Fork shales record the gradual encroachment of the sea over the land (marine transgression) which began some 380 million years before present (BP). The overlying, coarse-grained Kanayut was shed from northern uplifts which caused the seas to retreat (regression). After the mountains were finally eroded, the ocean, once again, transgressed this area and deposited the Kayak Shale. The total thickness of these rocks is approximately 14,000 feet!

Along the Highway, the Hunt Fork is mostly gray and brown in contrast to the black slate and phyllite of the Beaucoup Formation. The Hunt Fork Shale is mostly slate, with some brown-weathering siltstone and sandstone. Upsection, the Hunt Fork changes from deep marine to shallow marine sedimentary rocks and grades into the coarse-grained and quartz-rich rocks of the Kanayut Conglomerate. Within the Kanayut, there are shoreline deposits called the Noatak Sandstone, which are overlain by conglomerates, sands and silts deposited by braided and meandering streams. Red-colored Kayak shale and silts record the drowning of the Kanayut delta and floodplain after the mountains were weathered away.

Regionally, the Hunt Fork may grade downsection into the thick slate section of the easterly derived Beaucoup Formation. Faulting, however, obscures the nature of this contact. Similar faulting also affects the contact between the Kanayut Conglomerate and the overlying Kayak Shale.

In the western and central Brooks Range, the Hunt Fork and Kanayut typically make up a structural or tectonic allochthon which may be thought of as a rather large sheet of rocks which has been moved (as a unit) laterally great distances by thrust faults. In the Doonerak area, west of about mile 230, one of these Hunt Fork-Kanayut allochthons is thrust, tens of miles, over the top of a succession of the younger, middle Ellsmerian sequence (upper Mississippian through Triassic) rocks that are exposed in an erosional window (fenster). At Doonerak, these middle Ellsemerian rocks overlie argillite just as they are typically found in the subsurface of the North Slope or along the northernmost exposures of the Brooks Range.

The Endicott Mountains rise to the north and extend 150 miles to the west, forming part of the Brooks Range. The Brooks Range, with elevations ranging from 4,000 to 9,000 feet, extends from Canada on the east to the Chukchi Sea off western Alaska, a distance of about 600 miles. The range is named after Alaskan geologist Alfred H. Brooks, chief Alaskan geologist for the U. S. Geological Survey from 1903 until his death in 1924.

Mile 239

The Hunt Fork Shale and various members of the Kanayut Conglomerate compose most of the bedrock along both sides of the highway to about mile 267.

Mile 239.5

Chandalar Camp is a former pipeline construction camp. Climatic observations conducted by the U.S. Army Cold Regions Research and Engineering Laboratory from 1975-1978 show that higher elevations along the Dalton Highway have warmer average winter temperatures than lower elevation sites such as Coldfoot, Dietrich, or Prospect. The extremely cold temperatures (down to -80° F) recorded at these sites are caused by the movement of cold air downslope.

Cooler summer temperatures at higher elevations reflect the normal decrease of temperatures with elevation.

Mile 240

Possible open-system pingo on the west side of the road.

Mile 240.6

Here the highway traverses the bottom section of a steep **debris cone** composed of talus, mudflow deposits, and snow-avalanche debris. Alluvial fans are distinguished from cones by their gentle slope and braided-stream channels.

Does geology "bug" you?

Researchers from the Department of Geology at North Dakota State University found the northernmost known occurrence of a species of beetle at the Chandalar Shelf. These beetles were previously known only from the fossil record of glacial stages in Iowa. A discovery such as this demonstrates that even though the styles of glaciation differ between the Chandalar Shelf and Iowa, habitats and their species were or are similar. This suggests that some of the plant and animal life in the Brooks Range could provide clues to deciphering the environment that existed in the vicinity of the last great glaciation in the United States.

Leffingwell: Serendipitous Explorer-Famous Geologist

An unsuccessful attempt at geographic discovery, resulted in a geologic discovery of some significance. In the spring of 1906 a small expedition named the Anglo-American Polar Expedition was organized by Capt. Ejnar Mikkelsen and Ernest de K. Leffingwell to explore the Beaufort Sea, north of Alaska as far east as Herschel Island. Their purpose was to find a land mass north of the Arctic Coast in the Beaufort Sea.

Heavy ice conditions in the summer of 1906 badly damaged their ship, the Duchess of Bedford, in the ice 60 miles east of Prudhoe Bay. Leffingwell and crew salvaged the lumber from the "Duchess" and built their cabin on Flaxman Island. A long disappointing sled trip in the spring of 1907 did not find the land mass. The sled trip and ocean soundings determined that land could not exist to the north. Most of the crew boarded the first whaling ship and returned home. Mikkelsen decided to return alone by dog sled to Valdez via the coastline to Barrow, Kotzebue, and Nome, then overland to Fairbanks and Valdez, altogether a distance of about 2200 miles.

Supported by his father's money, Leffingwell decided to stay in Alaska. He remaine in the arctic for most of seven years. During this period, Leffingwell completed the first geology studies and permafrost studies in the Canning-Jago Rivers area in what is now the Arctic National Wildlife Refuge. He also surveyed the arctic coast of Alaska from Demarcation Point to Point Barrow.

Leffingwell's work was published in 1919 as USGS Professional Paper 109 and is considered a classic in geological literature. This work still stands today, as it was Leffingwell who first named the Sadlerochit Formation which in recent years proved to be the famous "pay sand" of the Prudhoe Bay Field. Leffingwell lived to see the culmination of work he started. He died in 1971 at the age of 96, three years after the discovery at Prudhoe Bay.

Figure 47. View southward of Chandalar Shelf. The pipeline is built on a landslide lobe overlying glacial till of Itkillik age. Conspicuous mounds of angular blocks on the lobe indicate that it may have been deposited on a stagnant and melting glacier.

Mile 241.3

A 12-foot-thick landslide lobe overlies glacial till on the valley floor east of the Chandalar River (fig 47). The till is of Itkillik age and was eroded from Devonian conglomerate, siltstone, and shale bedrock which form the mountains in this area. Conspicuous mounds of angular blocks on the landslide indicate that it may have been deposited on a stagnant and melting glacier.

Mile 242.5

The road crosses the west fork of the North Fork of the Chandalar River and begins the climb to Atigun Pass. Fossils of brachiopod and horn coral may be found in the silty **limestone** beds of Hunt Fork Shale located upstream on the south side of the fork.

The orange-weathering oxidized zones, visible on exposed sections of the Kanayut Conglomerate, are sometimes good indicators of significant mineralization (that is, a potentially valuable deposit). In this case, however, it is the oxidation of pyrite, a common iron and sulfur-bearing mineral.

Mile 243.8

Turnout near waterfall. A nearly complete section of Kanayut Conglomerate is exposed up the gully on the west side of the road.

Mile 244.5

Atigun Pass (elev. 4,643 feet). The continental divide (see fig. 53) separates river systems which flow to opposite sides of a continent. In this case, summits of the Brooks Range separate streams draining north toward the Arctic Ocean from streams flowing south (via the Koyukuk and Yukon rivers) toward the Bering Sea. (see also fig. 60)

Atigun Pass also marks the boundary between continuous (to the north) and discontinuous (to the south) permafrost along the Dalton Highway. From this point northward, permafrost thickness increases and is present almost everywhere including the ground beneath small streams and ponds. In discontinuous permafrost southward, frozen areas become less and less abundant and permafrost thickness decreases.

Contrary to what one might think, almost all areas underlying active glaciers are believed to be free of permafrost. The great thickness of glacier ice acts as insulation and interrupts the penetration of colder air temperatures at the ground surface.

A cirque carved out of the interbedded sands, slates, and conglomerate of the Kanayut Conglomerate is visible east of the road. Over a period of many centuries, the great weight of accumulating glacial ice cut into the bedrock and formed the armchair-like depression.

Cirque glaciers are relatively small, and are most common during the early and late stages of major glacial periods. During peak glacial stages, cirque glaciers often become part of a continuous body of glacier ice that extends over mountainous terrain or high plateaus. Mountain cap ice eventually gives rise to valley glaciers that carve large U-shaped troughs as ice moves downslope through the mountains. Atigun Valley is one such trough.

Mile 246

View to the north into Atigun Valley (fig. 48). The steep mountains and ridge crests in the region to the north are composed of weather-resistant conglomerate and sandstone of the Kanayut Conglomerate. This 8,600-foot-thick sequence extends some 600 miles across the east-west trending Brooks Range. It is one of the thickest and most widespread nonmarine sequences in North America.

The Kanayut was deposited about 350 million years ago in a delta-like environment characterized by a vast, braided plain and an extensive flood plain. Southwest-flowing

USDI BLM photo / William Diel

Figure 48. View northward of the glacially scoured Atigun Valley. Several lobate-shaped rock glaciers have formed in the Hunt Fork Shale just below taluses on both the east and west sides of the valley. The ice matrix within a rock glacier lets the rocky mass flow downslope, similar to the movement of an ice glacier.

Alyeska Pipeline Service Co. / Steve McCutcheon

Figure 49. Aerial view southeastward of Atigun Valley and mountain slopes composed of the Hunt Fork Shale. Note the rock glaciers on the lower right side of photo and that the highway traverses an alluvial fan.

USDI BLM photo / Arthur Banet, Jr.

Figure 50. View of a slushflow across the Dalton Highway near Atigun Pass. A slushflow is a type of wet-snow avalanche initiated by rapid spring melting of the seasonal snow cover. Slushflow deposits consist of unsorted rock fragments, large blocks, fine-grained material, and occasional shredded plant remains.

Slushflows are considered a natural hazard in northern alpine regions of Alaska and are common in stream courses of the Atigun Pass area. Here snow avalanches strip the surrounding bedrock loosened by frost action and deposit it in tributary valleys. Rapid snowmelt during mid-May to mid-June flushes out the debris and eventually forms an alluvial fan.

braided and meandering fresh-water streams out of a highland area to the northeast deposited rock fragments (mostly chert and quartz derived from preexisting rocks) in a lowland basin bordering a shallow sea.

Several lobe-shaped, ice-cemented masses of rocky rubble, called rock glaciers, have formed below taluses on the east and west sides of the glacially scoured, U-shaped Atigun Valley. The ice matrix within a rock glacier lets the rocky mass flow downslope, similar to the way an ice glacier moves.

Highway construction during 1974 exposed considerable ground ice in the fractures and joints of the bedrock on the west side of the road. Substantial roadcut degradation occurred the following summer. Because of the high-ice content of the bedrock in this area, about 800 feet of pipeline was buried in an insulated box on both the north and south slopes of Atigun Pass.

Mile 246 - 252

Between MP 246 and MP 252, the Hunt Fork Shale forms the bedrock slopes on both the east and west sides of Atigun Valley

Mile 247

A rock glacier and an alluvial fan have formed on the east side of the road. Numerous alpine fans and cones built by various fluvial and **colluvial** processes are visible along the lower slopes of Atigun valley.

The highway traverses an alluvial fan deposit that originates in the valley to the east. The fan, fed by an alpine stream that carries rock debris, bears scattered boulders strewn across its surface—an indication of periodic slushflow activity.

Steep talus cones, formed of debris deposited by rock falls and snow avalanches also provide debris to the lobate rock glaciers.

Mile 247.1

Highway crosses Atigun River. The road is built on an alluvial fan created primarily by slushflows.

Alluvial fans contain many boulders, and exhibit steep slopes that lack an extensive surface network of **braided streams**. Lobate rock glacier to the west.

Mile 247.7

Lobate rock glacier and debris cone to the east.

Mile 248.6

Roadcut through Hunt Fork Shale. Between here and MP 252 the Hunt Fork Shale forms most of the slopes on both the east and the west sides of Atigun Valley (fig. 51). A large thrust fault has placed the older (and originally underlying) Hunt Fork Shale over the younger Kanayut Conglomerate.

Mile 250

Atigun Camp. Roadcut in alluvial fan and debris flow to the east (fig. 52).

Mile 252

View to the south of Grizzly glacier.

Mile 252.5

The curved bedrock of the Hunt Fork Shale visible up the valley to the east are isoclinal folds. Folds are indicative of the great compressional forces that acted upon the once flat-lying shale and siltstone bedrock.

Mile 253

Atigun River crossing. Lobate rock glacier to the west of the road.

Mile 253.5

Several small rock glaciers along the western slope.

USDI BLM photo / Arthur Banet, Jr.

Figure 51. View of a synclinal fold in the Devonian Hunt Fork Shale. The Hunt Fork forms most of the slopes on both the east and west sides of Atigun Valley between Mileposts 247 and 253.

USDI BLM photo / William Diel

Figure 52. The combination of steep slopes, lack of vegetation and saturated sediments from melting snow causes numerous debris flows to form annually on the lower slopes of Atigun Valley.

Refer to figure 53 for geology and points of interest between Mile 255 and Mile 348.

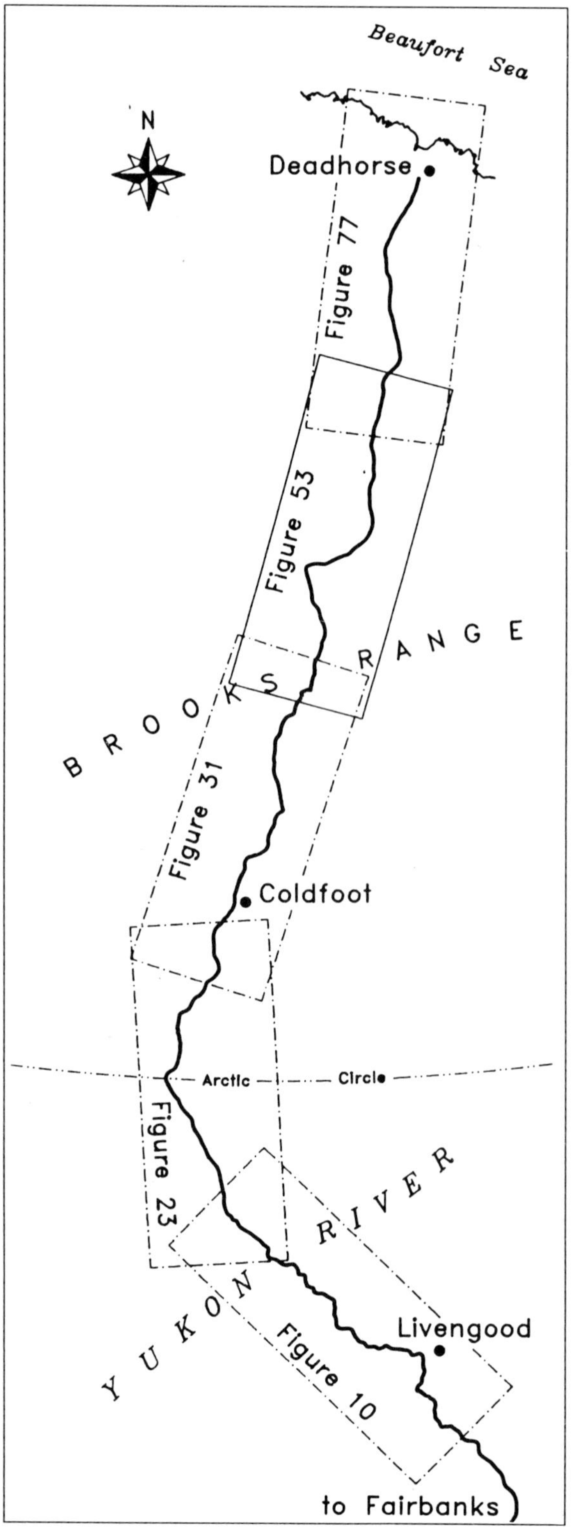

Mile 254.5

Road crosses **alluvial fan** deposit. Fan material originally deposited in the Atigun River flood plain has subsequently been washed away leaving behind large boulders of possible slushflow origin.

To the east, an east-west trending thrust fault places the darker colored Hunt Fork Shale on the south against the brownish Kanayut Conglomerate on the north.

View of a cirque to the west. Rock glacier on the south face.

High on the valley wall to the west is an exposure of the possible depositional contact between the severely faulted and folded Kanayut and the tightly-folded black Kayak Shale of the North Slope succession.

Mile 255

Pipeline emerges from underground.

Mile 257.5

View to south of James Dalton Mountain. This mountain is composed of Hunt Fork Shale and on the south is thrust faulted against the shallow-marine sandstone and conglomerate of Noatak Sandstone, on the north (fig. 54).

View to the southwest of the West Fork Atigun River valley. The geologic relationships seen in this view illustrate the differences between the relatively tightly folded and faulted Hunt Fork-Kanayut and the North Slope succession, which is a more open-fold style. Because of these different tectonic styles and depositional relationships, we separate the Hunt Fork-Kanayut and North Slope successions into separate terranes.

Mile 259.3

There is an overturned **syncline** in the Kanayut along the east side of the valley.

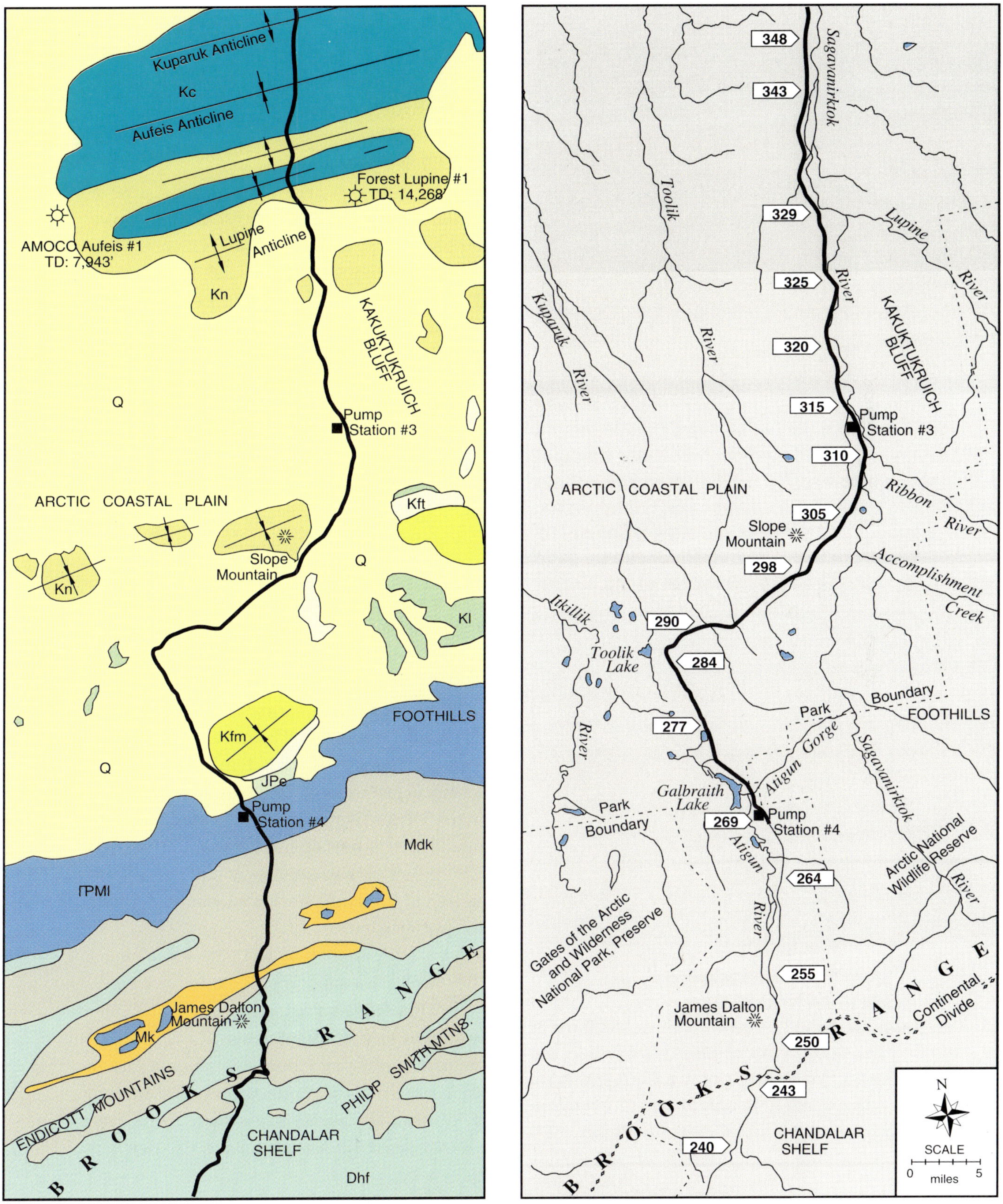

Figure 53. Dalton Highway geology and points of interest. Mile 255 to Mile 348

from Brosgé and others (1979), Dillon and others (1979) (1986) and Mull and others (1987). See figure 92 for key.

Alyeska Pipeline Service Co. / Steve McCutcheon

Figure 54. Aerial view looking west across Atigun Valley of a cirque carved mostly from the weather resistant Kanayut Conglomerate. On the northern edge of the cirque lies the contact between the Kanayut and the younger Kayak Shale. The West Fork of the Atigun River can be seen in the valley just north of the cirque.

Alyeska Pipeline Service Co. / Steve McCutcheon

Figure 55. View eastward across Atigun Valley of a large section of tightly folded Kanayut Conglomerate. The rock quarry in the bottom left of photo lies within the Shainan Lake Member of the Kanayut. Note the small fold composed of the dark-colored Kayak Shale near the upper right corner of the photo.

Mile 260

Gravel-based, braided channel of the Atigun River to west.

Mile 260.8

A road to a quarry with an exposure of braided stream deposits of the Shainin Lake Member of the Kanayut. **Matrix**-supported fragments are mostly composed of chert and other fragments of previously formed rocks. Dall sheep are often seen in this quarry.

Mile 264.1

Water from alluvial valley drainage melts the underlying ice-rich permafrost and erodes a small gully on east side of the road. Erosion is migrating up slope and away from the roadbed.

The Atigun River changes from a gravel-based, braided channel to a meandering channel composed of finer-grained lake sediments left behind by a large, ancient, **glacial lake**.

Mile 266

Kanayut Conglomerate forms the castle-like ridges to the west. The dark colored shale layers within the Kanayut interfinger with the more prominent and weather resistant conglomerate. Tight and overturned folds are common in this part of the Hunt Fork - Kanayut outcrop belt. These are the northernmost exposures of the Hunt Fork and Kanayut along the Highway.

Mile 266.5

In the valley to the east, enormous compressional forces squeezed the Kanayut Conglomerate into a syncline, a U-shaped fold in which the strata or limbs dip inward toward the center. The same forces have formed an arch-shaped fold or anticline to the north of the syncline.

North Slope Succession/Arctic Coastal Plain Sequence

Compressional faulting, which is illustrated so well in the Doonerak area, built the Brooks Range and has telescoped all Ellesmerian units (fig. 56). This repetition, or thrust faulting, of units on top of each other is called crustal shortening. Movement along these faults is determined by mapping the various thrust faults or allochthons. After the thrust faults and plates exposed on the surface are mapped, the faults and fault plates for the subsurface are reconstructed. Note that the faults and wild folds we see along the highway in thrust plates occurred at depth and have been uplifted by additional later faults. All these faults are then restored as nearly as possible to their original unfaulted and unrepeated position in what is called a palinspastic reconstruction. Thus, geologists estimate that several hundred kilometers of crustal shortening has occurred in the Brooks Range.

The contact between the North Slope Succession and the underlying Devonian clastic rocks is difficult to describe. This is one reason why they are treated as separate terranes. The Kanayut is overlain by sandstones and shales indicating a marine transgression of Mississippian age. The Devonian Hunt Fork and Kanayut were deposited in an elongate basin located far to the south, much further south than most (if not all) of the middle Ellesmerian part of the North Slope succession. Faulting has deformed and juxtaposed both the lower and middle Ellesmerian units in a complicated geometry. (Some excellent illustrations of this complexity are west of the highway between mile 225 and the mountain front.)

Also, neither the Hunt Fork nor the Kanayut are known north of the mountain front. Available drill hole and outcrop data, even from around Doonerak, typically show an upper Mississippian Kayak shale unconformably overlying lower Paleozoic (or older) argillite rather than the Kanayut or Hunt Fork. This is an enigma. The Kayak shale may be a single unit which is both the top part of the Kanayut transgression and the basal part of the middle Ellesmerian regression. Or, it may be that there are two similar but separate Kayak shales. The great distances between lower and middle Ellesmerian depositional centers and the differences in their ages suggests that either interpretation has possibilities. At this time, the Kayak relationship is even more enigmatic than the Beaucoup-Hunt Fork contact. As yet, neither mapping nor seismic data are available to reconcile whether this enigma is due to erosion, or faulting, some of both, or neither!

So, for simplification, the North Slope succession consists of lower Ellesmerian Hunt Fork Shale and Kanayut. A large **unconformity**, indicative of mountain building, separates these from the North Slope Succession consisting of middle and upper Ellesmerian sequence rocks. This took place at approximately 355 Ma. Both the central Brooks Range Succession and the North Slope Succession were shed from a northern landmass. The north Slope Succession, however, includes the basal, lower Paleozoic (and perhaps some older) shales, argillites or carbonates that had previously been uplifted and tilted. The middle Ellesmerian rocks are the upper Mississippian Endicott Group, Kayak Shale, Kekiktuk Conglomerate and the Mississippian-Pennsylvanian Lisburne Group limestones. These rocks record the erosion of a region-wide uplift. This erosion was immediately followed by an environment that had relatively warm stable conditions during which the Lisburne limestones were deposited.

Another unconformity from regional uplift separates the middle and upper Ellesmerian sequences. This took place at approximately 275 Ma. The Sadlerochit Group (Echooka, Kavik Shale and Ivishak) sands record the last major mountain building event. These sediments are coarse-grained delta and flood plain deposits in the subsurface along the coast, such as around Prudhoe Bay. They become thin bedded and fine grained toward the mountain front, where these distal sediments are mapped as Siksikpuk and Otuk Formations. Thus, most of what we know about them is from well data rather than outcrop data. The bulk of the well data is from the Prudhoe Bay area, the site of North America's largest oil field. The top of the upper Ellesmerian consists of thin sandstones and some organic-rich limestones and shales.

A: Hatching shows flat–bedded, thick–bedded rocks and the solid lines represent the thin and interbedded (less competent) rocks before deformation.

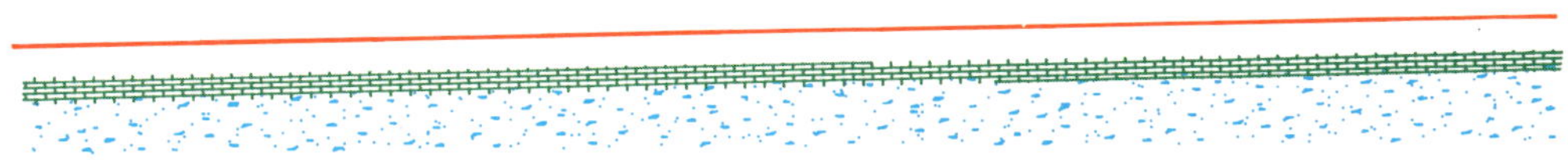

B: As compression begins, folds develop within the thin, less competent beds. Faults (shown as heavy lines) form in competent beds, first along bedding planes then upward through the thicker rock layers. Younger faults deform older faults and folds.

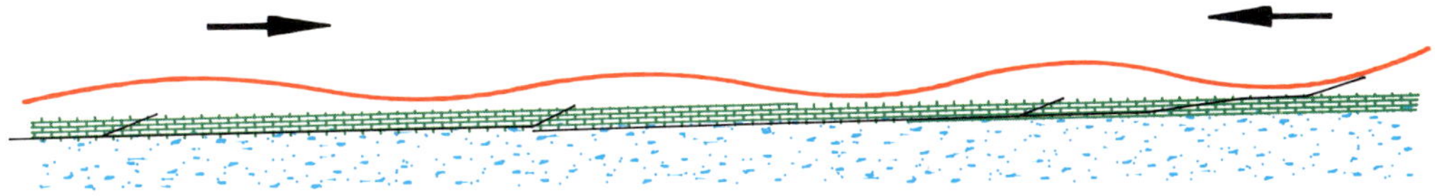

C: Continued compression pushes older rocks over younger rocks. The thin, incompetent beds are wildly deformed and overturned. The deep, thick, competent beds stack up like leaning books. Fault #1 uplifts strata and is subsequently uplifted by faults #2 and #3.

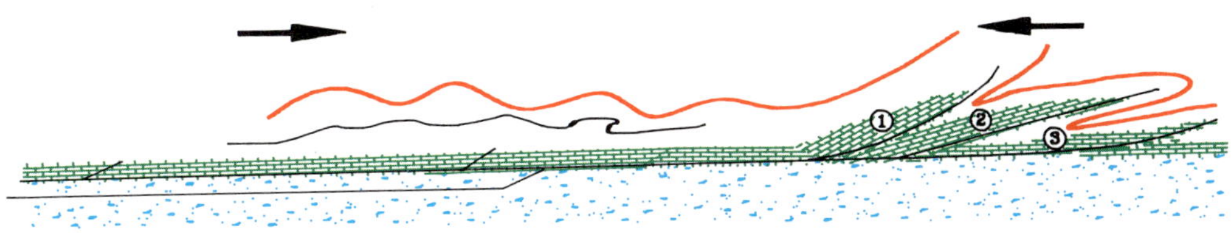

D: Faulting affects the deepest rocks, deforming all of the previously folded and faulted beds. Numbers indicate probable sequence.

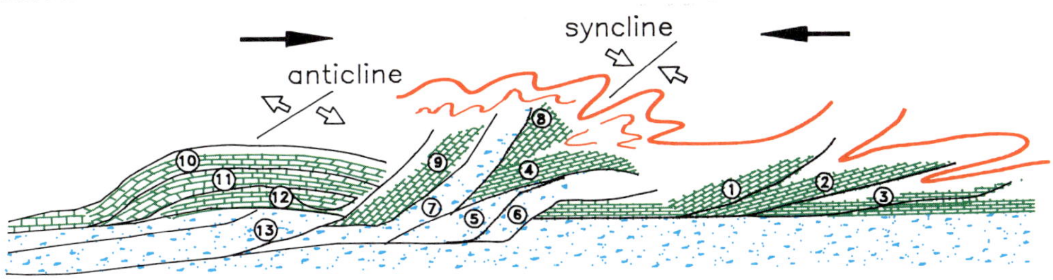

Thus, mountains are uplifted as rock layers are moved laterally many miles. Both anticlines and synclines are formed, folded and overturned as deformation progressively affects lower and more competent beds.

Figure 56. An idealized representation of how tectonic compression forms, folds and thrust faults in the kinds of rocks found along the Dalton Highway.

USDI BLM photo / William Diel

Figure 57. Pump Station 4 is built on a till-covered, glacier-scoured hill of Lisburne Group limestone. It is the only pump station north of Fairbanks built on an unrefrigerated foundation.

Alyeska Pipeline Service Co. / Steve McCutcheon

Figure 58. Aerial view northwestward of the broad, glaciated Atigun River Valley which has an irregular topography consisting of many developing thermokarst ponds, pits, and lakes. Valley fill consists of glacial and glaciofluvial deposits overlain by lacustrine sediments left behind by a fresh water lake that may have reached 18 miles in length.

Mile 268

Roadcut in alluvial fan material.

Mile 269.3

Pump Station 4 is built on a till-covered, glacier-scoured hill of Lisburne Group limestone (fig. 57). It is the only pump station north of Fairbanks built on an unrefrigerated foundation.

The Carboniferous Lisburne Group, which forms the spectacular, massive gray cliffs east of the highway, represents a carbonate rock sequence deposited in a shallow-marine environment.

The broad, glaciated Atigun River Valley has an irregular topography that consists of many developing thermokarst ponds, pits, and lakes (fig. 58). Valley fill consists of glacial and **glaciofluvial** deposits overlain by lacustrine sediments left behind by a fresh water lake that may have reached 18 miles in length (fig. 61). This large glacial lake was dammed by an end moraine located north of Galbraith Lake and subsequently drained through Atigun Gorge.

The Atigun River bedload consists of sand and flood-plain deposits, overlying fine-grained lake sediments that include silt and clay. Linear, sharp-crested ridges of silty sand called yardangs are visible along the margins of the Atigun River flood plain (fig. 59). Yardangs are carved by wind erosion and parallel the dominant wind direction, in this case, southeast to northwest. The associated troughs separating these **eolian** features may reach 70 feet across and 15 feet deep.

Mile 269.5

Road passes over pipeline. Note the absence of fin-like radiators atop the pipeline's vertical support members. These heat dissipating thermal devices are not needed because the ice-rich permafrost in this area re-

Figure 59. Aerial view northwestward of drainage of Galbraith Lake into Atigun River. Note the yardangs in the foreground. Yardangs are linear, sharp-crested ridges of silty sand carved by wind erosion. Troughs separating these eolian features may reach 70 feet across and 15 feet deep.

Figure 60. Color infrared aerial view of Atigun Pass and vicinity.

USDI BLM

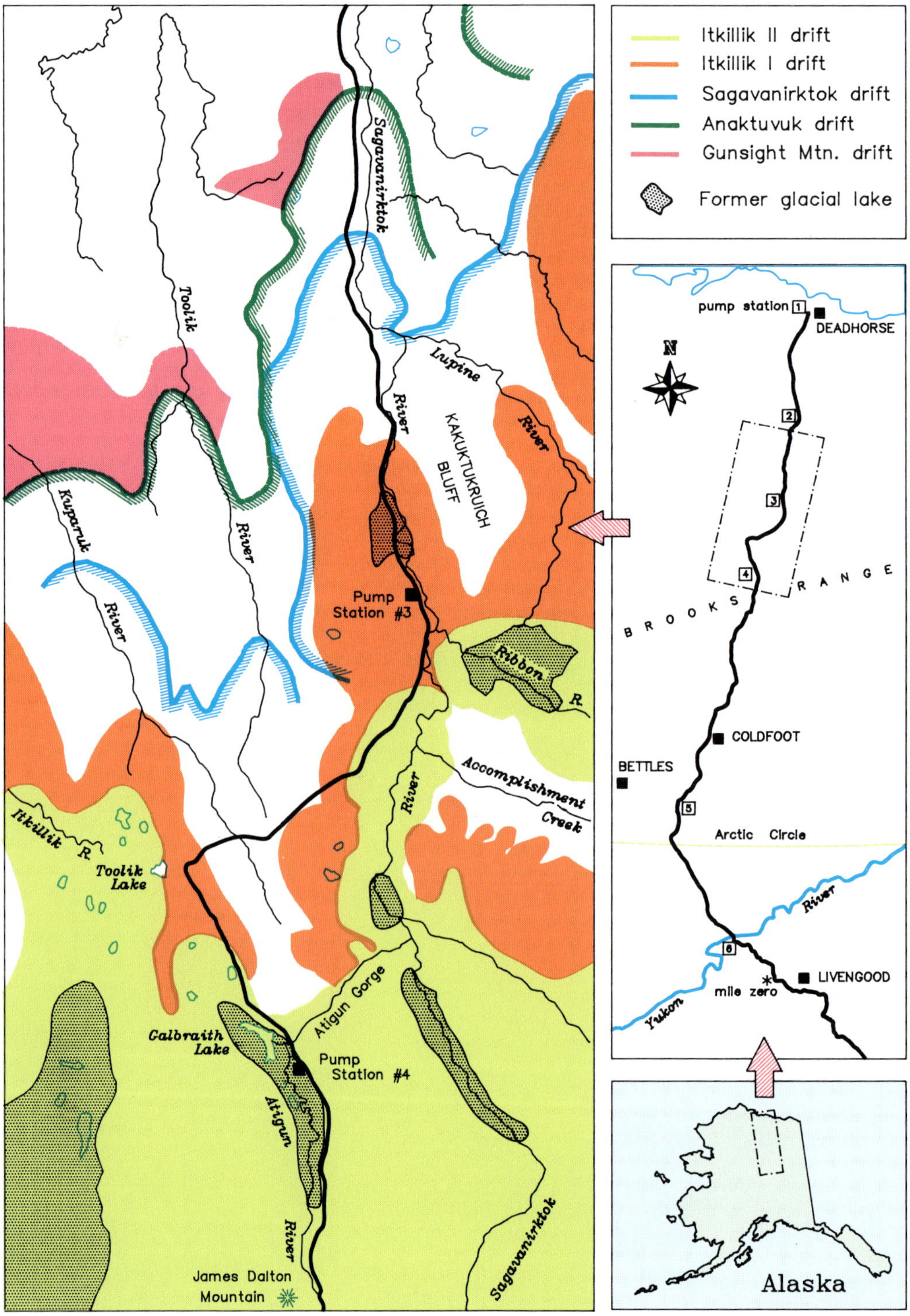

Figure 61. Glacial limits of the area between James Dalton Mountain and Sagwon Bluff.
(after Hamilton, 1978, 1979)

mains sufficiently cold at the base of the VSM's to prevent thawing.

Mile 270

Cirque in Lisburne Group limestone to the east. Silt/sand deposits to the west.

A day hike to the east up to the castle-like exposure of limestone takes you across three different types of structures in the Lisburne Limestone (fig. 62). The lower ridge is an anticline within the Lisburne. This is overlain by the black shales of the Permian Echooka Formation. A second sequence of Lisburne forms the ridgeline. This Lisburne is completely overturned and thrust-faulted over the lower sequences. The second sequence weathers to tan and consists of a ridge forming unit and a darker, thin-bedded unit which illustrates the folding. The mountain top consists of gray, south-dipping Lisburne thrust over the two lower units.

The underlying Kayak Shale and the overlying shales of the Echooka Formation, which normally occur in the stratigraphy, act as glide planes for these thrust-faults and do not form good outcrops. Where exposed, these shales are highly tectonized; that is, they are stretched into thin units or tightly folded into relatively small spaces.

Brachiopods, **bryozoans** and small corals can be recovered from the stratigraphic top of the tan-weathering, middle limestone unit (fig. 63). (Since this unit has been overturned, look for the fossils at the base of the unit.)

The view from the top of the castle-like ridge is both scenically and geologically stunning! To the south is the flat-bottomed, glaciated Atigun River valley. Note the looping river **oxbows** and the small, rounded thermokarst lakes.

There are unconsolidated glacial sands and silts which have been stripped of protective vegetation and eroded by the wind into "blow-outs."

Edward Bovy

Figure 62. View eastward of three sequences of Lisburne Group limestone located on the south side of Atigun Gorge. 1) The lower ridge forms an anticline within the Lisburne. This is overlain by the black shales of the Permian Echooka Formation. 2) A second sequence of Lisburne, which is completely overturned and thrust faulted over the lower sequence, forms the ridgeline above the Echooka Formation. 3) The mountain top above the thrust consists of gray, south-dipping Lisburne thrust over the two lower units.

USDI BLM photo / Arthur Banet, Jr.

Figure 63. Brachiopods (pictured), bryozoans and small corals can be recovered from the stratigraphic top of the overturned tan-weathering, middle limestone unit of the Lisburne Group.

Brookian Sequences

The Brooks Range was built by three major periods of mountain building caused by compressional tectonics (fig. 64). Small, widely isolated outcrops of mostly volcanic rocks suggest that the initial stages of Brooks Range mountain building may have begun in the upper Jurassic. This changed the basin polarity; that is, sediments now entered the basin from the south rather than from the north.

The lower Brookian deposits of the Okpikruak (Ko), and the Kongakut (Kk) Formations record the first deposits of deep water sandstones and shales derived from the south. By middle Cretaceous time, the ancestral Brooks Range was uplifted in earnest. Exceedingly thick deposits of interbedded sandstones, shale and conglomerates were shed into a trough that developed parallel to the present mountain front. These are the lower Brookian sequence Fortress Mountain (Kfm), Torok (Kt), and Nanushuk Group (Kn) which may be as much as 20,000 feet thick in the center of the tough. They filled the Colville Trough from the south-southwest to the east-northeast and together represent a mirror image of the older Hunt Fork and Kanayut deposition. Limited exploration drilling has thus far yielded only minor oil and gas discoveries from lower Brookian rocks.

After a hiatus of several million years during which only a thin veneer of marine shale was deposited, another pulse of mountain building began. Starting at about 94 (ma), up to about 15,000 feet of middle Brookian sediments were shed northeastward from the central and eastern parts of the Brooks Range. Along the Highway, these rocks are exposed in the cores of several synclines on the Arctic Coastal Plain. These are the Upper Cretaceous to Lower Tertiary Colville Group, which is divided between the lower, mostly marine Schrader Bluff (Kcs) and the nonmarine Prince Creek Formations (Kcp). Immediately west, and upsection of the Prudhoe Bay area, middle Brookian rocks are the **reservoir rocks** for large reserves of low temperature and low gravity oil. Unfortunately, the extraction of these oil deposits has eluded even the most current recovery techniques to date. Smaller, nearly economic discoveries have been found beneath the Arctic Coastal Plain, immediately outside the Arctic National Wildlife Refuge, 1002 area (fig. 88).

The final pulse of Brookian mountain building began in the late Eocene. Mostly nonmarine sediments were shed into central and northeast Alaska. Onshore, these are up to 8,500 feet of unconsolidated, coarse-grained sands and conglomerates. Offshore sediments are up to 20,000 feet thick. So far, four important discoveries have been made in the Brookian sediments north of the 1002 area.

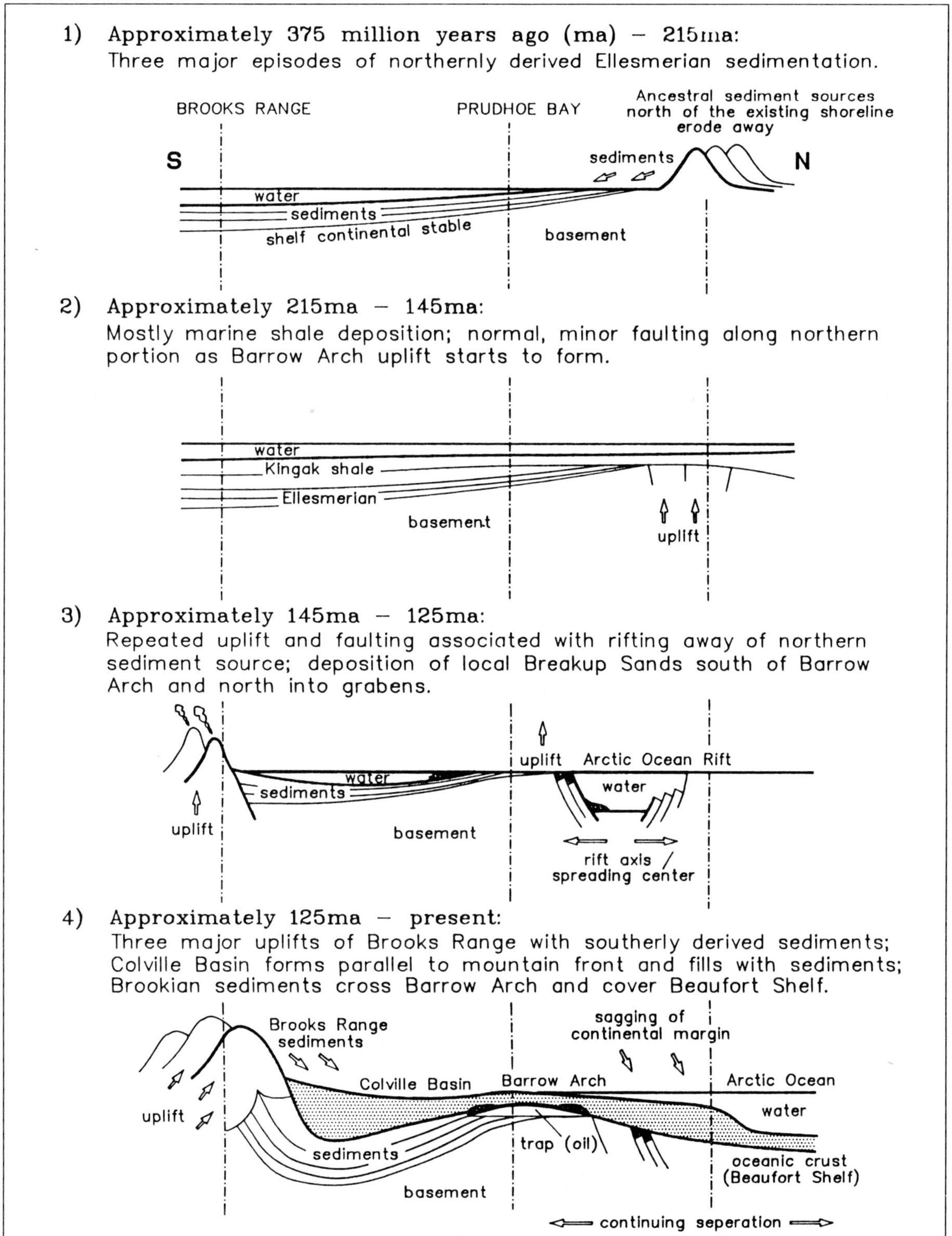

Figure 64. Geologic evolution of the North Slope, Alaska.
(adapted from Rickwood, 1970)

Above the valley, south dipping beds and folds within the dark-colored Hunt Fork and Kanayut are visible. Directly west, there are several folds and repeated sections in the limestone walls above Galbraith Lake. To the northwest, the mountains abruptly terminate into the relatively flat, glaciated portion of the North Slope. Atigun Gorge, to the northeast, shows gently folded Cretaceous sandstones and shales (Kfm) overlying steeply dipping and contorted upper Paleozoic and older Mesozoic shales.

Mile 270.5

Thermokarst pond on east side of road occupies a slightly higher elevation than the modern Atigun River channel.

Mile 270.9

Atigun River crossing. The Atigun River heads at a glacier **terminus** in the Endicott Mountains to the south, then flows northeast 45 miles to join the north-flowing Sagavanirktok River. Moderate to well-sorted sand is exposed in a road cut on the north side of Atigun River bridge. These sands were originally deposited as alluvium that filled the Atigun Valley basin. They now accumulate as dune sand that began to form more than 750 years ago and which continues to be reworked by strong down-valley winds.

Mile 271.5

Atigun Gorge to the east (fig. 65).

South-dipping, faulted Lisburne limestone and Echooka sedimentary rocks crop out on south side of Atigun Gorge. Virtually perpendicular and highly distorted upper Ellesmerian and Brookian sedimentary rocks crop out on the north side of Atigun River.

The view down Atigun Gorge illustrates a major change in the nature of North Slope deformation. To the south, the thick lime-

Figure 65. Aerial view eastward of Atigun Gorge. The gorge separates the 330-million-year-old Lisburne Group limestones on the south (pictured) from the 105-million-year-old, mostly sandstone and shale, Fortress Mountain Formation to the north. The view down Atigun Gorge illustrates a major change in the nature of North Slope deformation. (See Milepost 271.5 in text for further discussion.)

Edward Bovy

Figure 66. View of aufeis (icings) that form annually on an alluvial fan west of Galbraith Lake. Aufeis are thick masses of surface ice formed in winter by the successive overflow and subsequent freezing of sheets of surface water or emerging ground water. Note the shallowness of the water in the stream compared to the thick, successive layers of ice that have accumulated through the winter months.

stones and clastics of the Ellesmerian sequence in the Brooks Range have been regionally uplifted, folded, and thrust faulted. This compressional deformation has moved or telescoped these rocks from their original locations many miles to the south. To the north, Cretaceous sediments are typically broadly folded and are most visible in the axes of fault-bounded anticlines and synclines. The Cretaceous sediments are mostly sandstones and shales. These are also deformed by compression but they typically deform in a manner which uplifts them vertically more than laterally as with the older sequences.

Mile 272

Galbraith Lake, about 3.5 miles long and of glacial origin, was named for bush pilot Bart Galbraith in 1951.

Mile 272.5

Entering the Arctic Foothills physiographic province. View of a large icing (aufeis) that forms annually on an alluvial fan west of Galbraith Lake (fig. 66)

Mile 273

A thick sequence (9,000 feet) of conglomerate, sandstone, and shale of the Fortress Mountain Formation forms the mountain slopes north of Atigun Gorge (fig. 67). This sequence of rocks records a depositional history characterized by a rapid marine-to-continental transformation brought about by Brookian mountain building tectonics about 100 million years ago. These rocks are in the lower part of the Brookian depositional sequence marking the beginning of sediment deposition derived from the uplift of the ancestral Brooks Range to the south.

Mile 275.5

Numerous glacial erratic boulders, in glacial drift of Itkillik age, are strewn about the landscape west of the road (fig. 68). Erratics are carried by glacial ice and deposited at some distance from their source (in this case the Brooks Range).

Mile 277

The highway traverses a series of **recessional moraines** left behind by a melting and retreating Itkillik II glacier.

Mile 284.5

Toolik research camp of the University of Alaska Institute of Arctic Biology at Toolik Lake (to the west), has been the site of biological and botanical research for many years. Toolik Lake, a large kettle lake formed in glacial deposits of Itkillik II age, marks the former location of a large lingering block of stagnant glacial ice. A smaller kettle lake lies east of the road.

Figure 67. The Cretaceous Fortress Mountain Formation, consisting of conglomerate, sandstone, and shale, form the mountain slopes north of Atigun Gorge.

Figure 68. A glacial erratic boulder split in two by frost wedging.

Small hills west of Toolik Lake have steeply dipping Fortress Mountain and Kingak Shale bedrock. Larger hills to the north, Itigaknit, Imnaviat and Slope mountains, are Nanushuk Formation sandstones and shales preserved along a synclinal axis.

Mile 285

The road crosses the outer limit of a **lateral moraine** of Itkillik I age (fig. 61). Note the broad, rolling, tussock-covered ridges to the east. This is characteristic of the Sagavanirktok glacial stage which preceded the less extensive Itkillik stages. This landscape is markedly different from the steeper, drier, more irregular boulder-strewn moraines of the Itkillik stage.

The front of a double moraine enclosing Galbraith Lake is visible to the south.

Mile 286.1-298

The road crosses drift of Saganvirktok

Figure 69. Toolik Lake, a large kettle lake formed in glacial deposits of Itkillik II age, marks the former location of a large lingering block of stagnant glacial ice.

Edward Bovy

River age which is older and less pronounced than drift of Itkillik age.

Mile 288.9

To the northwest is an animal-crossing along the pipeline route. This short section of buried pipeline is insulated to prevent thawing of surrounding permafrost.

Mile 289.2

The pipeline crosses under the road through a large culvert.

Mile 290

Kuparuk River crossing.

Mile 297.5

The highway crosses the outer Itkillik I drift limit at the west margin of the Sagavanirktok River valley.

Mile 300

The road crosses the outer Itkillik II drift limit at the west margin of the Sagavanirktok River valley. Solifluction lobes are visible to the west.

Mile 301

This is the northernmost jurisdiction of the Bureau of Land Management along the pipeline corridor.

Slope Mountain to the west (fig. 70). The sequence of sedimentary rocks that forms Slope Mountain were deposited about 100 million years ago in an ancient marine to nonmarine, delta-like environment, where river derived sediments interacted with a larger body of water such as an ocean or gulf.

Marine shale beds of the Torok Formation, which form the lower-most section of Slope Mountain, were deposited just below wave action in what is called the prodelta zone.

Alyeska Pipeline Service Co. / Steve McCutcheon

Figure 70. View to the northwest of Slope Mountain, a sequence of sedimentary rocks deposited about 100 million years ago in an ancient marine to nonmarine, delta-like environment.

The Torok is overlain by two formations of the Nanusuk Group, the Tuktu and Chandler formations. Marine sandstone and shale of the Tuktu Formation formed in a delta front environment consisting of a continuous sheet of sand and occurring at the effective depth of wave erosion (about 30 feet or less). The delta front represents a narrow zone where deposition in deltas is most active.

The lower part of the Chandler Formation, which forms the weather resistant sandstone cliffs across the center of the face, represents the transition from marine to nonmarine deposition. These sandstone, shale, and siltstone beds were laid down inland as far as the zone of tidal influence (see interdistributary bay in fig. 71).

Lying above the transition zone are the nonmarine sandstone and conglomerate of the upper part of the Chandler Formation. These nonmarine deposits were laid down in the delta plain, an area above the tidal zone dominated by riverine processes.

Across Sagavanirktok River to the southeast, rock outcrops in the hills consist of alternating beds of sandstone and shale of the Fortress Mountain Formation and the dark graywacke sands and black pebbly shales of the Kongakut Formation. Hillsides further up Accomplishment Creek are composed of dark shales and graywacke units of the Fortress Mountain Formation.

The Gallagher flint station is one of the oldest archaeological sites in northern Alaska. It occupies a large **kame** north of the highway. Radiocarbon dating of material excavated in the early 1970's shows that this location was used by man as much as 10,000 years ago with sporadic usage as recent as 3,000 years ago. Cultural artifacts exhibit influences of both Inuit and Aleut people.

The Strange Disappearance of Sir John Franklin

Sir John Franklin, of the British Royal Navy, was an English Arctic explorer. Starting from the Mackenzie River in Canada, he explored the north coast of Alaska in July–August, 1826. He planned to rendezvous with another explorer, F.W. Beechey at Pt. Barrow. Bad weather turned him back and he did not complete the exploration.

In 1847, Franklin and his crew disappeared while on another expedition to explore and map the Arctic. It was the several search expeditions sent out to find Franklin in 1849-1854 that finally completed the work that Franklin set out to do. Many geographic features were named by Franklin or in honor of him and his crew by the search expeditions.

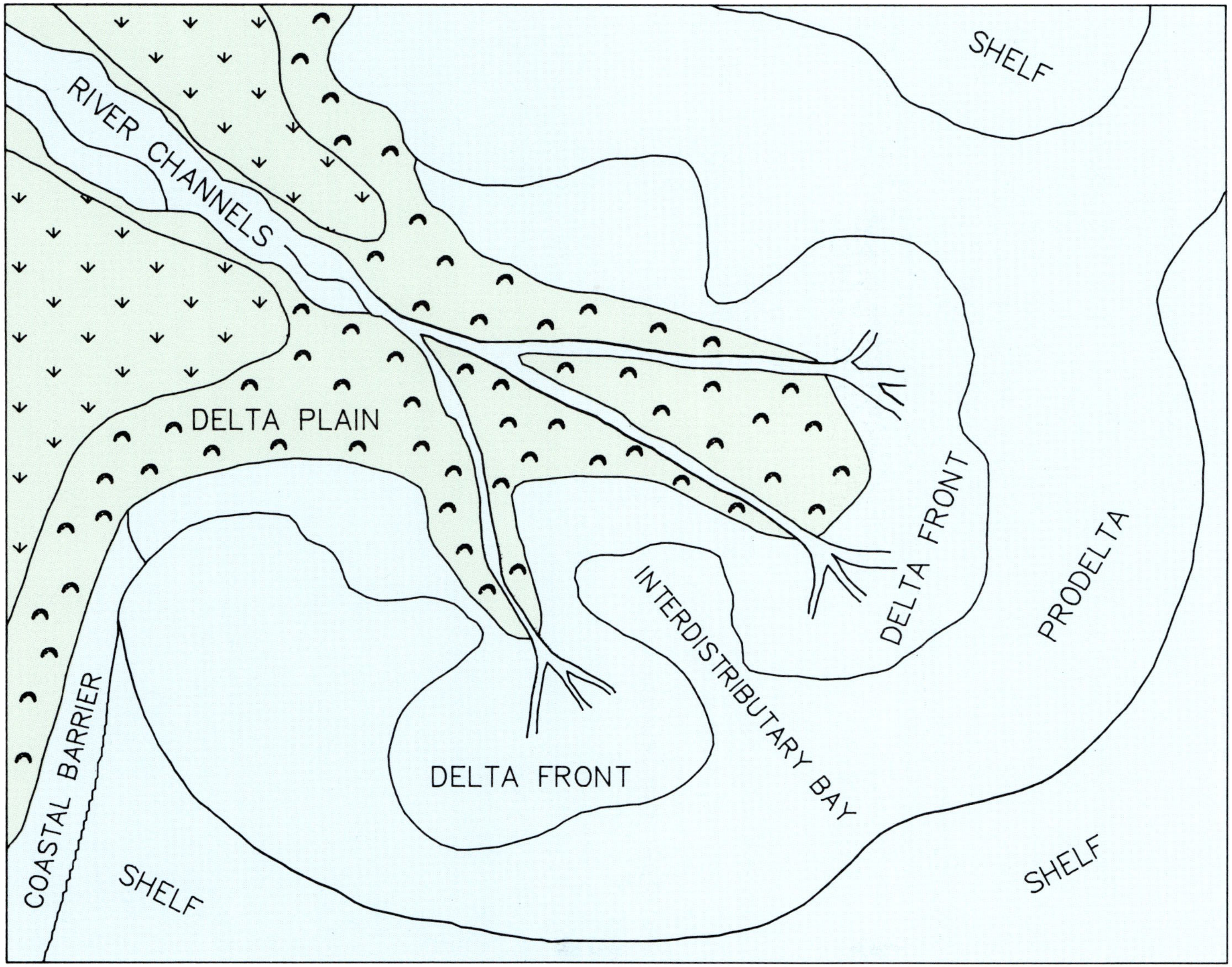

Figure 71. Deltaic depostional environments.
(modified from Huffman and others, 1985)

Breakup Sequence

Deep ocean conditions across the North Slope followed the final phase of sedimentation from the north. Over two thousand feet of Kingak Shale (JKk) was deposited during the Jurassic and early Cretaceous. In the subsurface, along the present-day coast, well-data record, minor pulses of various lower Cretaceous sedimentation events from several local uplifts. These uplifts formed sporadically during the **rift**ing, which separated the continental type rocks which were the source of the Ellesmerian sequences from the North Slope. These uplifts occurred between 210 Ma to as recently as, perhaps, 95 Ma, with the most regional events at 144 and 128 Ma.

Together, these uplifts are called the Barrow Arch. It dominates the subsurface geology for some three hundred miles, from the Barrow area to about the Canning River (fig. 89). On the south side of the Arch, the Ellesmerian rocks, and even some older rocks, were uplifted and tilted so that several thousand feet of sediments were sequentially eroded from along the crest of the Arch during the early Cretaceous. This erosion is primarily what created the subsurface geometry of the Prudhoe Bay field.

The sands and shales deposited from the uplifts are the Breakup sequence. They overlie the Ellesmerian and older rocks along the Arch and to the south. The Kuparuk Field, the second largest oil field in North America, as well as several more undeveloped fields are composed of these sands. To the north of the arch, Breakup sequence sands were deposited into deep troughs created by rifting. Exploration drilling has not yet been done to reveal the nature of these sediments.

Mile 305

Road crosses front of moraine of Itkillik II glaciation. The hillside west of the highway is built of moraines from both Sagavanirktok River and Itkillik drift.

Mile 305.6

Road to the right leads to a maintenance camp maintained by the Alaska Department of Transportation and Public Facilities. The camp does not provide gas or mechanical services to the public, however, a public phone is available.

Mile 310

Gully erosion on east side of road was induced by the concentration of water flow through culverts onto frozen, ice-rich soils where nonconcentrated flow previously existed. **Ground ice** is exposed on the north facing wall of the gully. Headward erosion toward the roadbed has apparently stopped.

Mile 313

Pump Station #3.

Mile 320 - 353

Brookian sequence bedrock crops out on hillsides. The sandstones are lower Brookian Nanushuk Group. These are overlain by middle Brookian shale of the Schrader Bluff Formation which is exposed in several east-trending synclinal axes. The northernmost part is composed of poorly consolidated sands of the upper Brookian Sagavanirktok Formation.

Mile 324-339

The road crosses **ground moraine** and end moraines of the Sagavanirktok River Glaciation.

Oops!

Sinclair Oil and Gas Company was an early industry promoter for oil exploration on the North Slope. The financially troubled company needed an "elephant" size discovery to bolster its dwindling reserves.

The company's exploration manager, Loren Ware, felt that the North Slope of Alaska was the best bet for a large discovery and in 1958 began the first steps in the exploration process.

During 1964 & 1965, the company had to choose between two upcoming sales, one in the Colville Delta, the other in the Prudhoe Bay area. It had seismic evidence of a large structure in both areas. The company chose the Colville Delta sale, as it was first.

The BP-Sinclair Colville # 1 (1966), penetrated a thick section of Permian-Triassic deltaic oil- stained sandstones-possibly the Sadlerochit. This gave the financially troubled oil company the first indications of a potential oil field on the North Slope. However, for financial reasons, and despite the efforts of Ware, Sinclair did not bid in the Prudhoe Bay area sale.

In 1967, Sinclair drilled another well on its Colville Delta lease. It was an uneconomic discovery. In 1969 Sinclair Oil and Gas Company was bought out by ARCO, one of the joint operators at Prudhoe Bay.

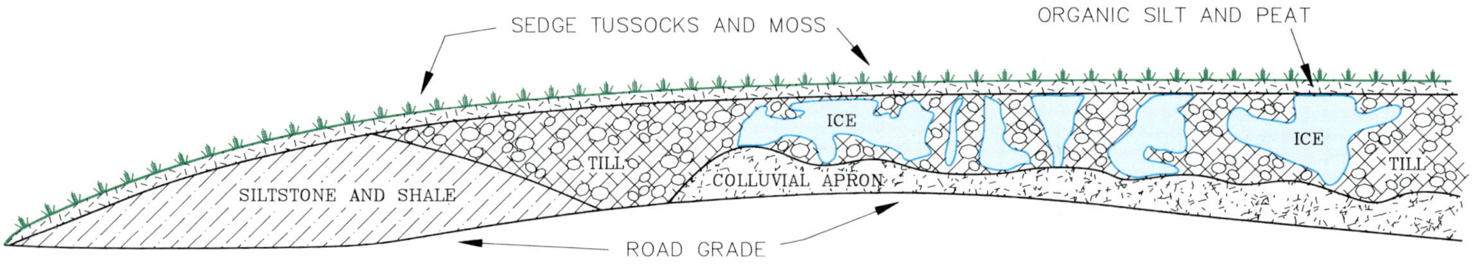

Figure 72. Happy Valley Roadcut
(after Kreig and Reger, 1982)

Figure 73. View northward of the braided flood plain of the Sagavanirktok River.

Mile 325

Happy Valley road cut. The road leaves the flood plain of the Sagavanirktok River and climbs a steep grade to a road cut exposing bedrock and glacial deposits (fig. 72). During highway construction in 1974, this section of the road exposed muck and peat with ice wedges and other massive ground ice lying above Sagavanirktok River Glaciation till resting on bedrock. By July 1975, massive ground ice was no longer visible in the cut, but continued melting produced slope retreat through slumping and flowage. This grade can be quite challenging to drivers whenever icy conditions exist.

This road cut also exposes nearly vertical dipping bedrock. The hard, bedded fine-grained sandstone or siltstone and black shale represent deep water deposits. They are the most distal portion of the Nanushuk Group or possibly part of the Torok Formation.

View to south of the northward-flowing Sagavanirktok River (fig. 73).

Mile 329

This is the approximate **axis** of the Lupine anticline (fig. 74). Note the steeply south dipping beds of Nanushuk Group rocks.

The Lupine Unit No. 1 oil exploration well was spudded, (to break ground by drilling) about 5 miles east of the road in 1975. The well was drilled to a depth of 14,268 ft. It had only minor gas accumulations and was subsequently abandoned. The mountains located about 30 miles to the east are made of Lisburne limestone and dark sandstone of the Echooka Formation.

Mile 343

Axis of the Kuparuk anticline. Note the gently south-dipping beds of the middle Brookian Schrader Bluff Formation visible east of the river.

Mile 344

Note scattered exposures of middle Brookian **bentonitic** Schrader Bluff Formation. Solifluction lobes are visible on the slope just west of road.

Mile 348

Axis of Ivishak anticline. Exposures of shale are of the middle Brookian, upper Cretaceous Schrader Bluff Formation. These are overlain by poorly consolidated sandstones and conglomerates of the Sagavanirktok Formation. The Cretaceous-Tertiary boundary at this location does not coincide with changes in rock types.

The Sagwon airstrip was a major staging camp for geological and geophysical parties in the 1960's. This also marks the northernmost extent of the Anaktuvuk River Glaciation.

Mile 350

Gravel quarry in hills.

Toe-may-toe or Tuh-ma-toe; Prew-doe or Pruhd-hoe?

The correct English pronunciation would of course be Prud (short u, rhyming with crud)- hoe Bay. In fact, this is how the name is pronounced by most locals of Prudhoe in Northumberland, England. A few of the English pronounce the village name as Prewd-hoe. Early explorationists followed the proper rules of pronunciation and used the short u, Prud-ho. This is the pronunciation listed in the Pronouncing Dictionary of British Names. The majority of Americans opt for the long u (Prewd-hoe)pronunciation. Why, is not known for sure, but perhaps this pronunciation comes from the pronunciation of news broadcasters who, in turn, probably adopted the southern drawl pronunciation of stateside oil workers.

Figure 74. General geologic diagram along the Dalton Highway from the Brooks Range to Prudhoe Bay.

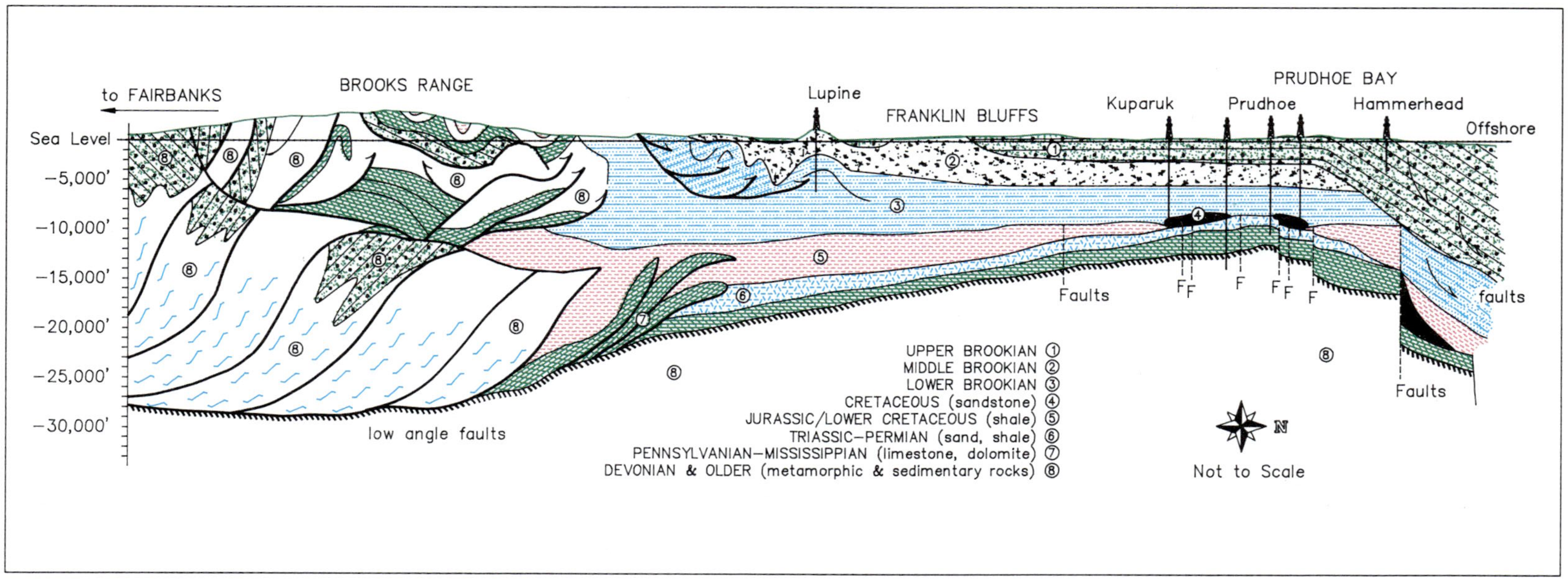

Geology and the dinosaurs

The Prince Creek Formation (see mile 352.3) is unique in that bones of Late Cretaceous (about 65 to 76 million years ago) dinosaurs have been recovered from its siltstone beds along the Colville River, about 90 air miles to the northwest. In all, at least six different dinosaur species have been identified at the site; hadrosaurs (duck-billed dinosaurs) are the most abundant. The fossilized hadrosaur bones are remarkably well preserved, displaying little premineralization, crushing or distortion. Bone length measurements indicate that these prehistoric creatures were 26 to 40 feet long, 10 feet high and weighed as much as three tons. They walked with their tails outstretched, not dragged behind.

Reconstruction of **plate tectonic** movement back to Cretaceous time tells us that all of northern Alaska (including the Prince Creek Formation) occupied a more northerly position that it does today. Even so, fossil plant debris found with the bones indicate a mild-to-cold temperate climate where wintertime temperatures seldom dipped below freezing.

At present, the Colville River discovery is the northernmost paleolatitude (latitude of the earth's surface in the geologic past) known for dinosaurs. Many questions remain unanswered. Why are amphibian and reptile fossils, commonly found with this kind of dinosaur, missing from the Colville River site? Did the hadrosaurs remain year-round at this northern latitude, or did they migrate south during winter months? If they did not migrate (the bones of juveniles found at the site suggest they didn't), how did they survive at latitudes where persistent darkness lasts several months, stopping plant growth?

USDI BLM photo / Robert King

USDI BLM photo / Robert King

Figure 75. (above) A cliff more than 120 feet tall has been exposed by the erosion of the Colville River.

Figure 76. Archaeologists carefully excavate a dinosaur bone.

Refer to figure 77 for geology and points of interest between Mile 348 and Mile 414.

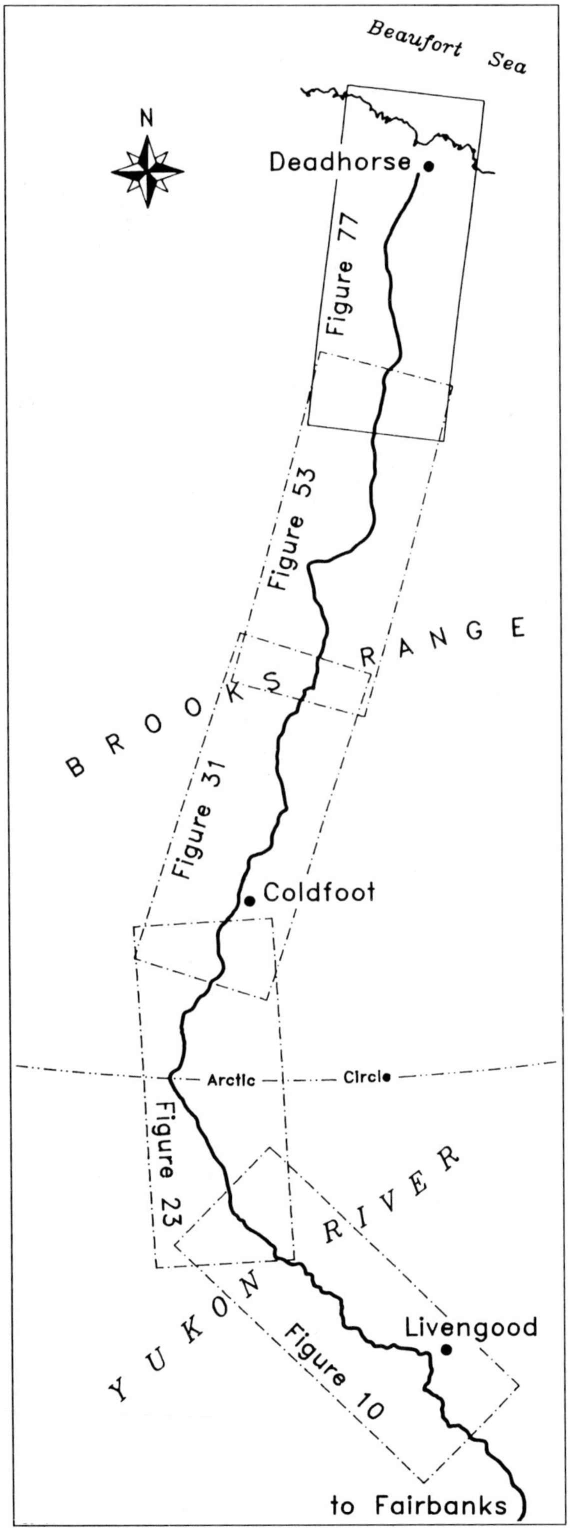

Mile 352.3

Weathered, easily-crumbled, and **crossbedded** fluvial sandstone of either the Cretaceous Prince Creek Formation or the Tertiary Sagavanirktok Formation crops out east of the road (fig. 78). This sandstone is very porous (contains voids or other openings) making it an excellent reservoir rock for oil and gas accumulations. The sandstone has similar physical characteristics and is about the same age as the sandstone in the Ugnu and West Sak oil fields located about 10 miles east of the Prudhoe Bay field. Nearly 60 percent of the world's oil and gas fields are in sandstone; the other 40 percent are divided mainly between limestone and dolomite. However, the total quantity of oil reserves is about the same in the two rock types.

Mile 354

Blading and compaction of the tussock-covered tundra by winter and summer tractor trains in the 1960's caused underlying ice wedges and ground ice in the fine-grained loess and colluvium of the Sagwon uplands to melt. A number of thermokarst ponds and pits developed as a result.

Mile 355

Turnout on hilltop. To the north lies the Arctic Coastal Plain physiographic province, an area crossed by many northward flowing streams and characterized by a network of **polygonal** ground and oriented thaw lakes. Most of the province is poorly drained because of the gentle northwest gradient and widespread, shallow permafrost.

Aerial photographs clearly show the large, ice wedge polygons (6-100 feet in diameter) that cover most of the province, including its drained lake basins and older fine-grained **alluvial** deposits (fig. 79).

Ice wedges are vertical, vein-like masses

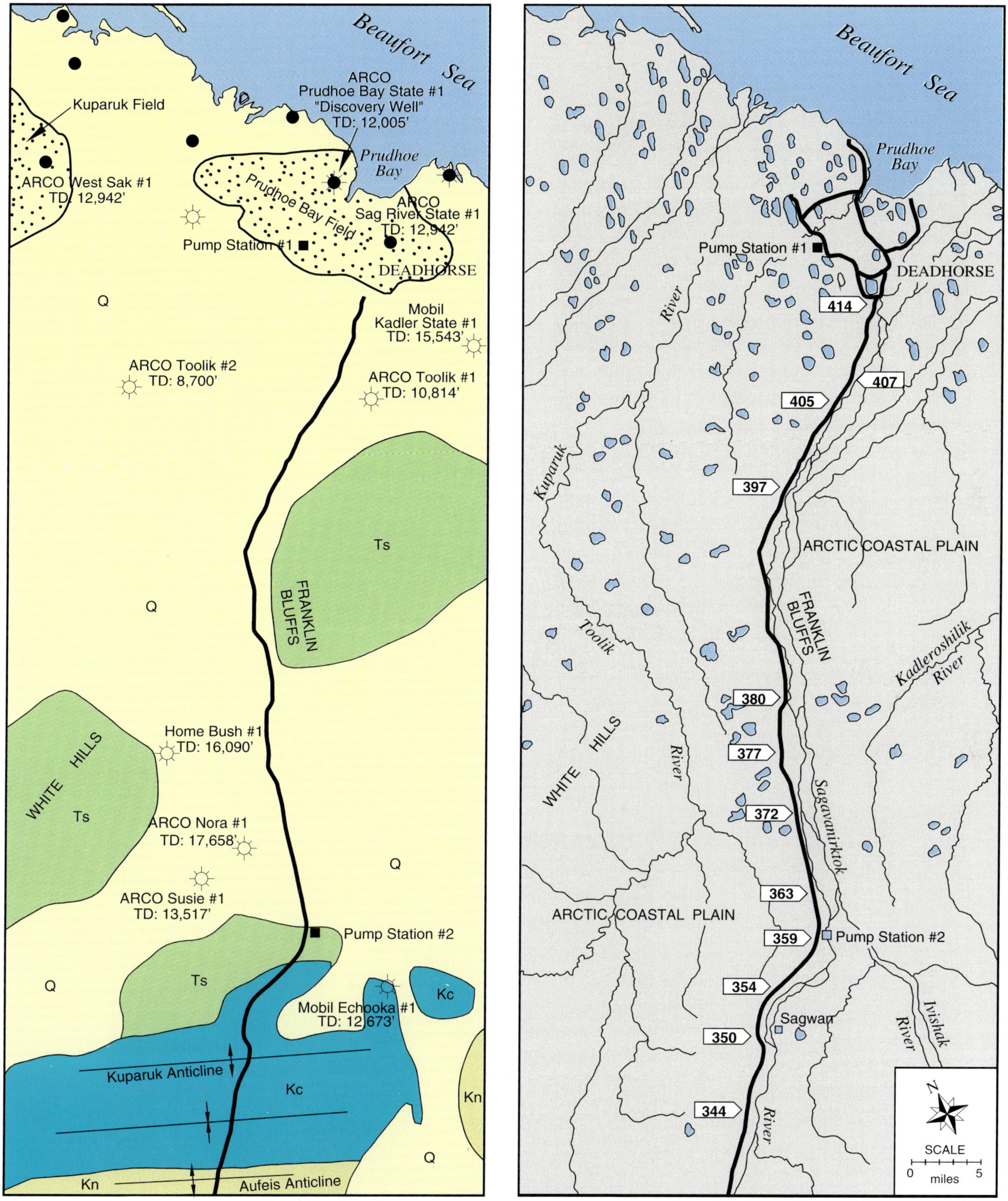

Figure 77. Dalton Highway geology and points of interest Mile 348 to 414

(From Mull and others (1987). See figure 92 for key.)

USDI BLM photo / Arthur Banet, Jr.

Figure 78. Weathered, easily crumbled, and crossbedded fluvial sandstone representing either the Cretaceous Prince Creek Formation or the Tertiary Sagavanirktok Formation near mile 352; it is very porous and an excellent reservoir rock for oil and gas accumulations.

Figure 79. Pump Station 1 at Prudhoe Bay marks the beginning of the trans-Alaska pipeline. Note the numerous polygon structures in the foreground.

of ice that develop by the accumulation and subsequent freezing of water vapor deposits (hoar) or melt water (fig. 80). They are part of a polygonal network of ice enclosing cells of frozen ground (the polygon) and form as a response to contraction and cracking of the ground surface as it cools rapidly. The large cracks extend vertically downward several feet and normally reopen and close annually. Repeated cracking at the same place over a period of years produces well-developed ice wedges.

Raised-edge polygons, or those having lower centers, indicate active ice wedge growth (fig. 83). Sediments are pushed up into ridges above and to the side of the underlying V-shaped ice wedge by continued deformation of the adjacent ground as it expands during the warm period. These slight surface depressions initiate the pooling of water and are frequently

Alyeska Pipeline Service Co.

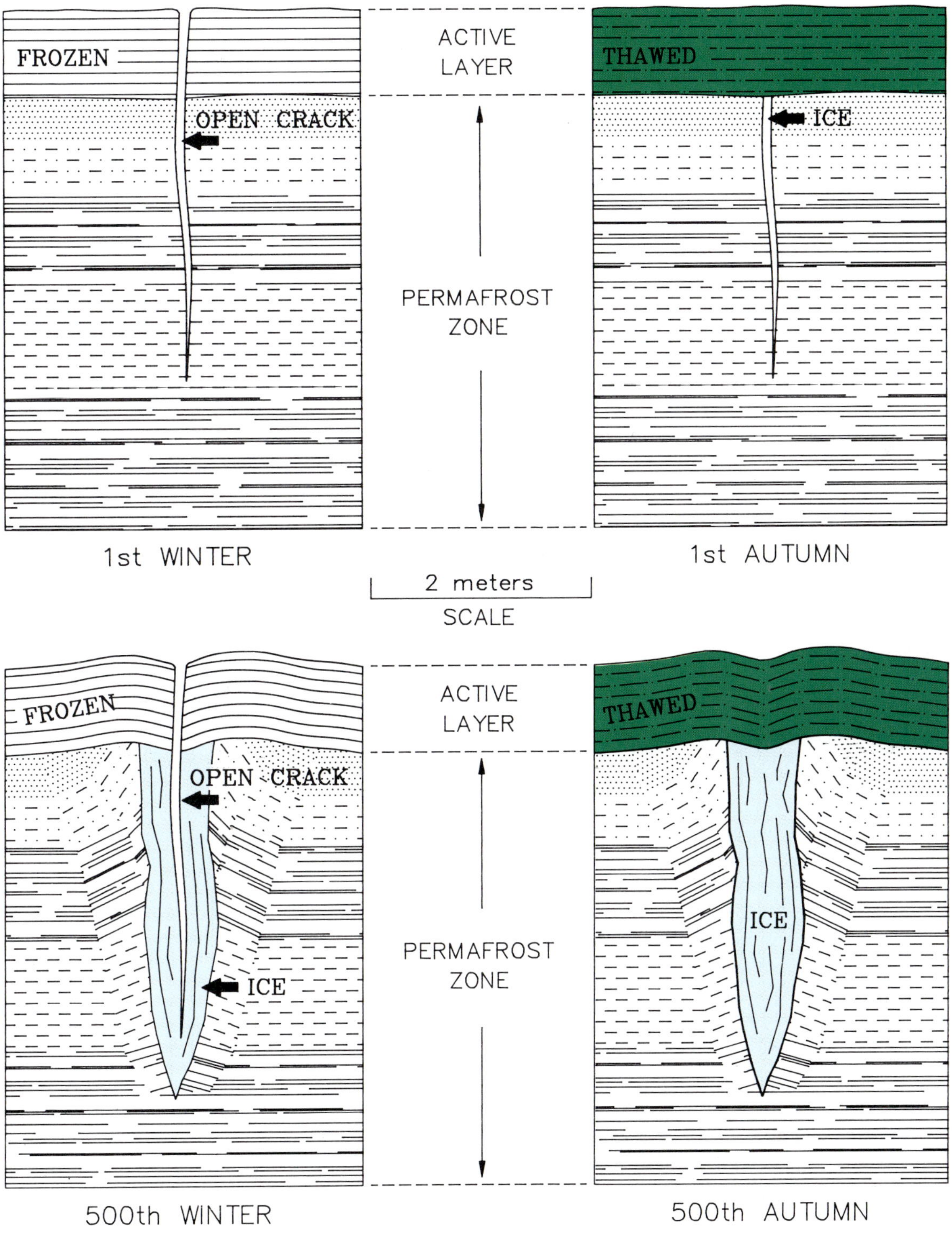

Figure 80. Evolution of an ice wedge.

Petroleum Formation

Petroleum molecules, first formed in the pores of organic-rich, fine-grained (smaller than sand) sedimentary materials, are squeezed out of the sediment pore space during deep burial. Buried sediments from which liquid or gaseous **hydrocarbons** are formed are called **source rock** (fig. 81). During expulsion from the source rock, petroleum will move or migrate along pathways of least resistance. One of the easiest migration routes is through coarse-grained sedimentary rock with connected pores such as sandstone. This migration, which may range from a few feet to tens of miles, continues until either the petroleum reaches the surface (in the form of seeps) or the oil and gas is trapped and can no longer move upward within the reservoir rock (fig. 82).

Hydrocarbon traps within reservoir rock are the ultimate targets of oil and gas exploration and drilling. Traps should not be visualized as huge voids within the earth's crust filled with oil and natural gas. In reality, hydrocarbons are located in small but abundant interconnected pore space within the reservoir rock. Pore space in a reservoir sandstone, for example, lies between the grains of sand.

Figure 81. Oil and gas formation requirements

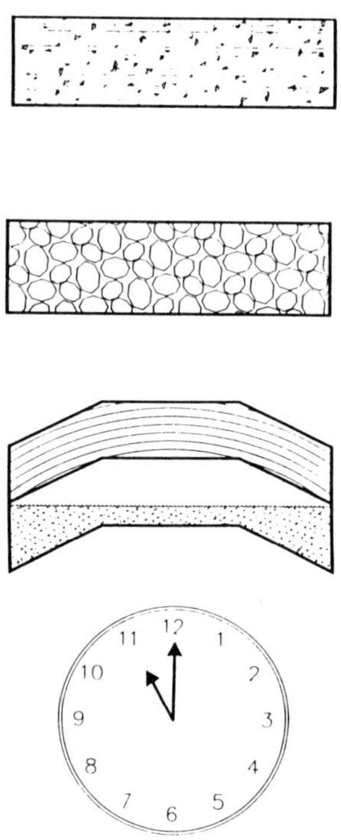

Source rock
An organic-rich rock, such as shale, buried at sufficient depth to "cook" and form oil.

Reservoir rock
A porous rock type with sufficient pore space for the oil to accumulate and "flow" or "migrate."

Trap
Deformed reservoir rock, sealed by an impervious cap rock, "traps" the oil from further migration.

Time
Events must occur in the proper sequence through millions of years. For example, a trap must be formed prior to the onset of oil migration.

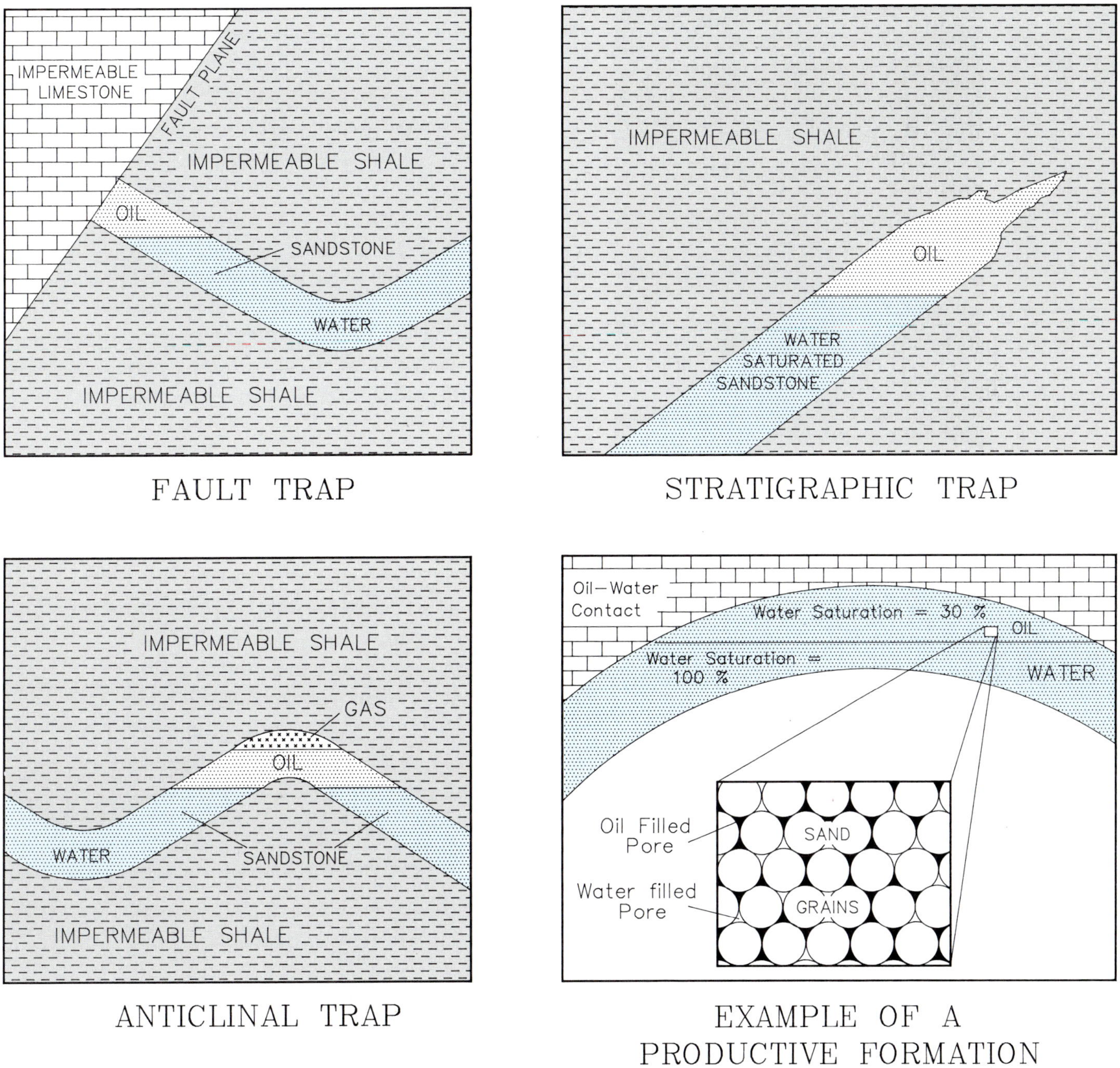

Figure 82. Types of hydrocarbon traps.

Figure 83. Aerial view of raised-edge polygons indicative of active ice wedge growth. Sediments are pushed up into ridges above and to the side of the underlying V-shaped ice wedge by the continued deformation of the adjacent ground as it expands during the summer period. These slight surface depressions initiate the pooling of water and are frequently the sites of thaw lake development.

the sites of thaw lake development.

High-center polygonal ground develops when the rate of deposition, thawing, or erosion is greater than the pushing up of sediments. They are normally found on slightly higher ground with better drainage. This leads to a faster thawing of the underlying ice wedges.

Innumerable northwest-oriented thaw lakes are the most notable features on aerial photographs or topographic maps of the Arctic Coastal Plain (fig. 84). Oriented thaw lakes may range from a few yards to several miles in length and are characterized by long axes perpendicular to dominant summer winds. They are constantly changing modern features that migrate, enlarge, drain, coalesce and re-form.

Lake formation begins with accelerated thawing of ice-rich soils due to a disruption in the thermal balance between permafrost and the tundra surface. Under the influence of the predominating winds, thermo-erosional niching undermines the banks which eventually collapse and recede.

The mechanism behind the formation of these lakes is still unresolved, but the most widely accepted view suggests that the lakes are oriented by winds blowing at right angles to the current predominant wind direction.

Mile 357

Good view of the east side of the Sagavanirktok River where basal Tertiary Sagavanirktok sandstones and conglomerates dip to the north under the Arctic coastal plain. Oil sand exposures, located upstream along the east side of the river, generated much speculation during the early days of oil explorations, before the Prudhoe Bay discovery.

Mile 359

Pump Station 2.

USDI Bureau of Land Management

Figure 84. Northwest-oriented lakes along the north slope develop perpendicular to the prevailing summer winds.

Mile 360-375

Only the drill pads remain of several exploration wells drilled several miles west of the road. In January, 1967, Atlantic Richfield and Humble Oil drilled the ARCO Susie #1 exploration well nearby. Although this well did not contain appreciable amounts of oil and gas, their next exploration well, the ARCO Prudhoe Bay State #1, drilled at Prudhoe Bay in 1967 and 1968, proved to be the discovery well of the largest oil field in North America.

Mile 363-405

Pipeline is buried.

Mile 372

White Hills to the west are exposures of the Sagavanirktok Formation.

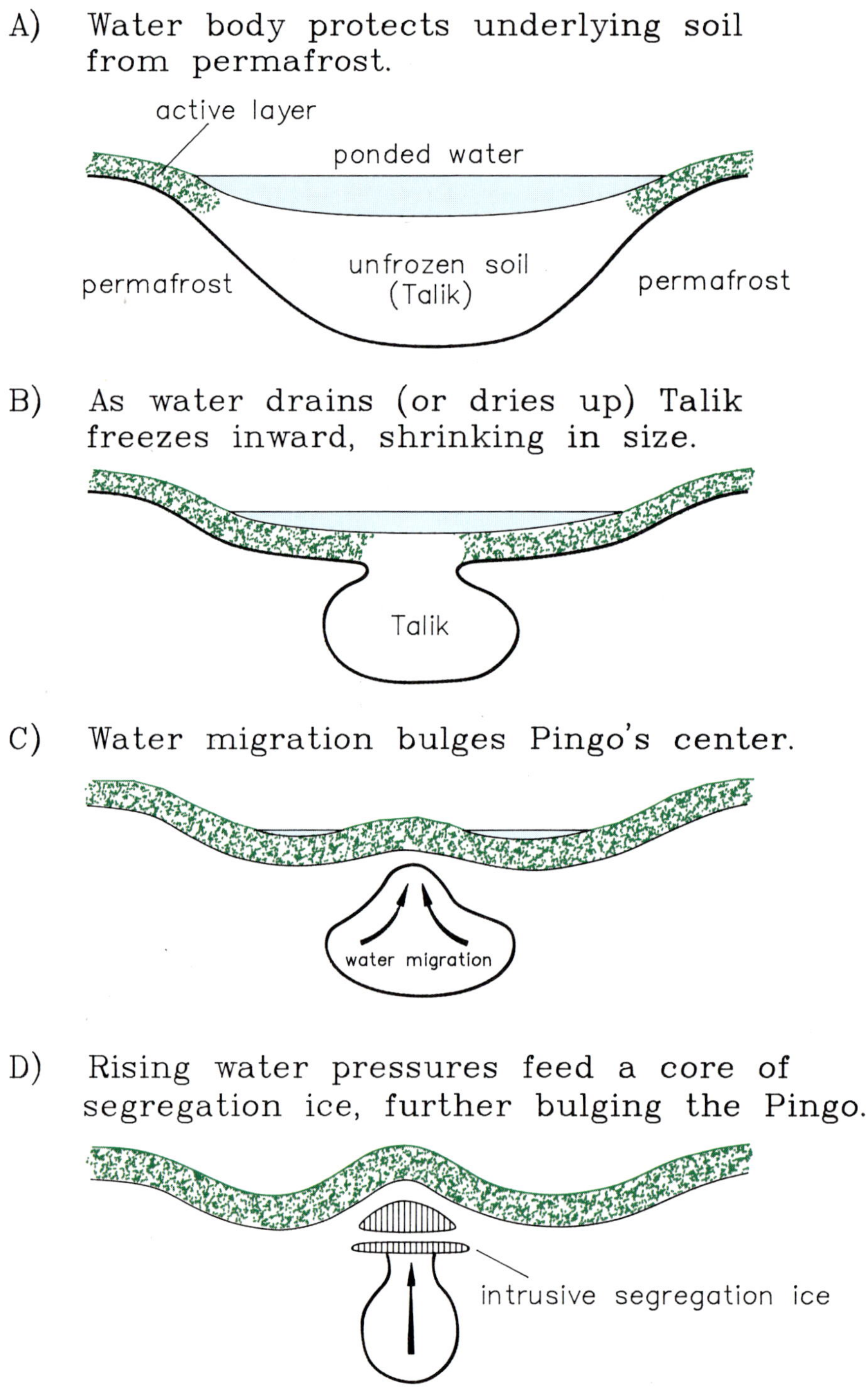

Figure 85. Closed system pingo formation.
(after Kudriavtsev, 1978)

Mile 377-397

Several closed-system pingos are visible in the distance to the west of the road.

Closed-system pingos (figs. 85 and 86) have developed in the refrozen, lake-basin sediments of the Arctic Coastal Plain physiographic province. This type of pingo forms in nearly level, low-lying, frost-free ground on drained or sediment- filled ponds or lakes. As the lake becomes shallower, the advancing permafrost forces pore-water out of coarse-grained sediments to form ice lenses, and frost heaving develops in fine-grained material. These processes combine to raise the ground surface, forming the pingo.

Figure 86. Pingo in the foothills of the Brooks Range.

Mile 380-394

Franklin Bluffs visible to the east. These bluffs display brightly colored yellow, tan, and orange clays and poorly consolidated sandstones and conglomerates of the Sagavanirktok Formation. Oxidation of iron minerals within these rocks gives them their distinctive colors.

Mile 382

Pingo to the west.

Mile 383-386

Gray, tan, and orange hues on Franklin Bluffs. Debris cones dissect bluff face.

Mile 405

Pipeline emerges.

Mile 407

Scalloped banks and tilted soil blocks produced by thermo-erosional niching can be seen on many sections of the Sagavanirktok River (fig. 87).

Thick river icings commonly form on many sections of the braided Sagavanirktok River flood plain. The shallow water channel, which freezes readily at the onset of winter, over-

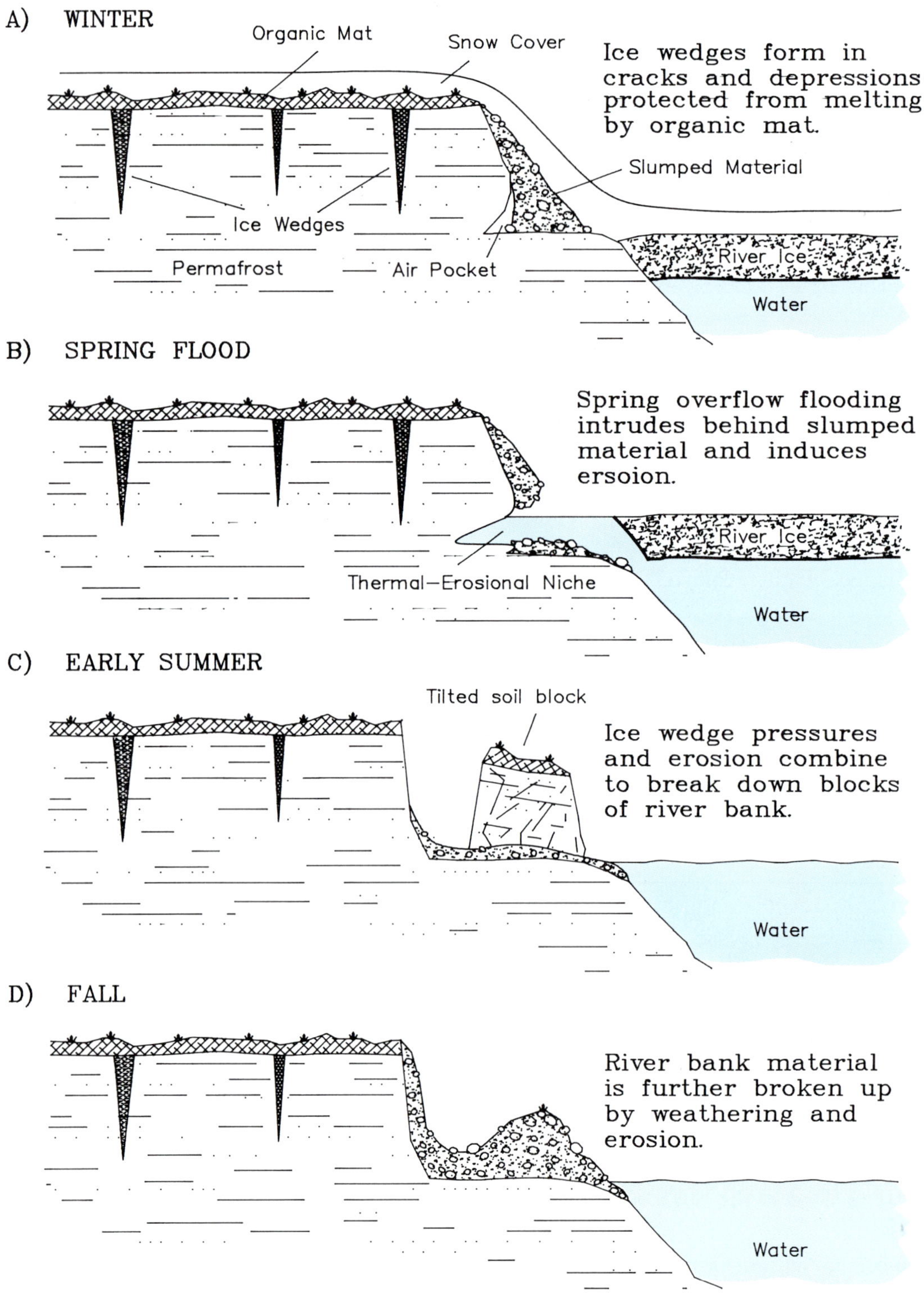

Figure 87. Thermo-erosional niching along the Sagavanirktok River.

flows its banks, often under hydrostatic pressure because of continued water flow beneath the ice. Icings are not common on large rivers that have a single or narrow, deep channel.

Mile 414

Deadhorse (Prudhoe Bay). The end of James Dalton Highway. Turn left to go to the airport.

More polygonal ground to the west. Wells drilled for petroleum reveal that the permafrost layer extends downward to about 2,100 feet in the Prudhoe Bay area.

East of the Prudhoe field, towards the Lisburne oil field development, there is a field of periglacial sand dunes. Strong seasonal winds commonly strip tundra vegetation from the dunes.

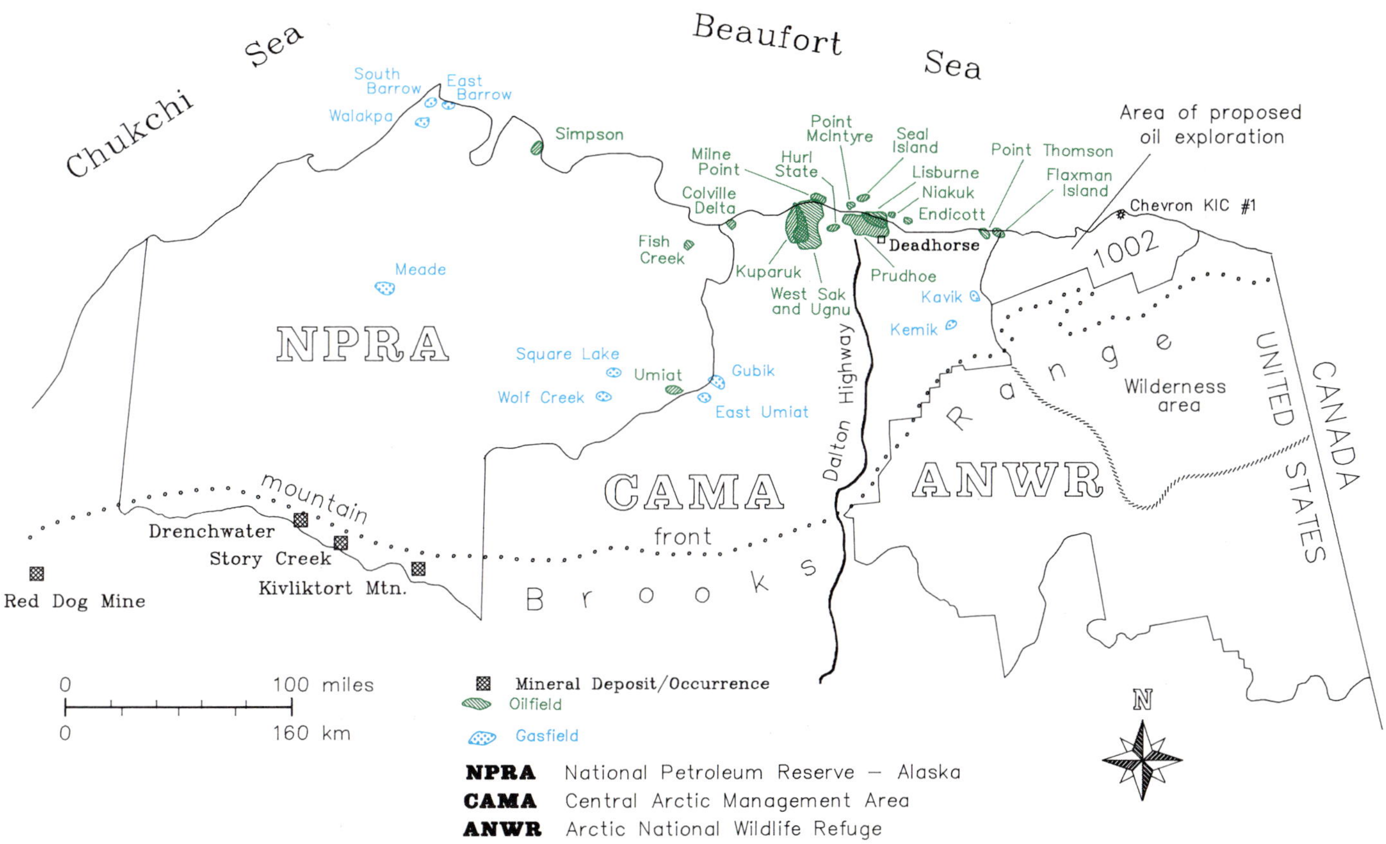

Figure 88. Oil and gas fields on the North Slope, Alaska.

Prudhoe Bay Oilfield

The Prudhoe Bay Oilfield, discovered in 1967, is the largest oilfield in North America. Upon discovery, it contained an estimated 20 billion barrels of oil-in-place with an estimated 9.6 billion barrels of recoverable oil. Enhanced oil recovery techniques have raised the estimate of recoverable oil to more than 10 billion barrels. The amount that the field will ultimately produce will be known only when production ceases. The ultimate production will depend upon several factors, including, but not necessarily limited to: the geology of the reservoir, the price of oil, improvements in recovery technology, and politics.

Geologically, the Prudhoe Bay Field lies in the midst of a group of closely spaced oil-rich reservoirs which lie atop a structural high in the basement rocks (the Barrow Arch) along Alaska's northern coast (fig. 89). These include the Permo-Triassic, the Lisburne Group, the Endicott, the Kuparuk River Formation, the West Sak sands, and the Ugnu sands reservoirs. The Permo-Triassic, Lisburne, and Endicott reservoirs are in the Ellesmerian Sequence, while the other reservoirs are in the Brookian Sequence (These sequences are discussed elsewhere in the road guide). The Prudhoe Bay Field lies within the Permo-Triassic reservoirs beneath an erosional surface (unconformity) at the base of the Brookian Sequence. The Permo-Triassic reservoirs include the Ivishak Sandstone of the Sadlerochit Group, the Shublik Formation, and the Sag River Formation.

The Ivishak Sandstone, deposited in an ancient river delta, contains most of the oil in the Prudhoe Bay Field. Two sequences of fine-grained to medium-grained pebbly sandstones are recognized within the Ivishak. In the lower sandstone sequence (about 300 feet thick), thin shale beds represent deposition in open bays between major channels of the ancient delta. The clean sandstones of the lower sequence are massive and, in places, conglomeratic. A conglomerate sequence (40 to 140 feet thick), which represents deposition in a non-marine environment, lies between the lower and upper sandstone sequences. These conglomerates mark the maximum southward extent of the Sadlerochit delta. The upper sandstone sequence (about 200 feet thick) consists of homogeneous, fine-grained to medium-grained sandstones. The lower part of the upper sandstone sequence was probably deposited in braided streams, and the upper part was probably deposited in a nearshore, marginal-marine environment.

The marine Shublik Formation, made of phosphate-rich, clay-rich, broken-shell limestones and calcium carbonate-bearing shales, overlies the Sadlerochit Group. The fine-grained, well-sorted, biologically-churned, marine Sag River Formation sandstones overlie the Shublik Formation. The erosional unconformity at the base of the Brookian Sequence played a major role in the formation of the reservoirs.

This unconformity exposed the rocks to subaerial (open-air) weathering which, according to some geologists, enhanced their porosity. Subsequent deposition of clay-rich sediments above the unconformity provided the seal necessary to trap the oil and gas that eventually migrated into the reservoirs.

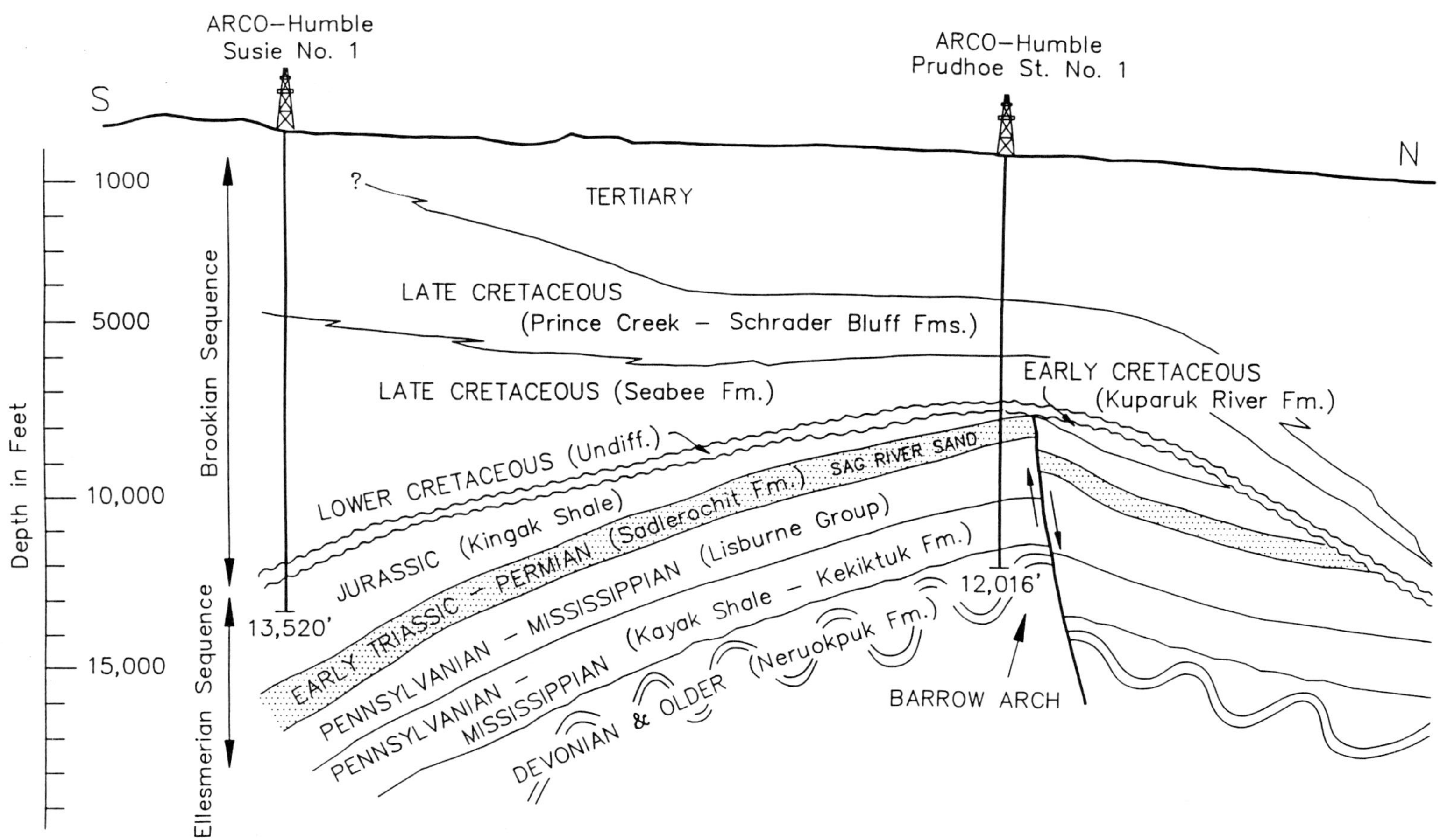

Figure 89. Regional cross section showing a stratigraphic section at Prudhoe Bay.
(modified from Specht and others, 1987)

ALASKA'S NORTH SLOPE

Alaska's northern frontier is the North Slope. In the early 1800s, the earliest explorers came to the arctic insearch of the Northwe st Passage. These early expeditions developed the first rudimentary maps of the Arctic Coast. During the decade before the civil war , the nation's need for oil brought the next generation of explorers to the area. They were looking for whale oil to fuel the nation's lamps! Increasing deman d brought the whalers to the Alaskan Arctic in search of the great sperm whale. As the demand for whale oil increased, it's supply dwindled and prices rose. This led to the search for alternative and less expensive fuels.

The discovery of petroleum in Ontario (1858) and Pennsylvania in 1859 was the answer. Although Eskimos had known of oil seeps and had burned oil shale long before the white man's arrival on the North Slope, it was not until 1886 that Ensign W. L. Howard of the U.S. Navy recorded what was probably oil shale along the Etivluk River in what is now The National Petroleum Reserve-Alaska (NPRA).

Despite this discovery, the North Slope was not explored or mapped until the turn of the century. In 1901, Frank Schrader and W.J. Peters of the USGS undertook the first systematic study of the geography and geology of the Brooks Ra nge and North Slope. Schraeder reported structures and rock types important to oil formation and entrapment. The USGS briefly noted the reported seep at Cape Simpson in Bulletin 379 in 1909. Between 1906 and 1914 and using private funds, Ernest de K. Leffingwell surveyed the arctic coastline between Barrow and t he Canadian Border. He also published a more complete description of the oil seep at Cape Simpson in his 1919 publication.

Government interests in the North Slope began in earnest in the 1920s following establishment of Naval Petroleum Reserve #4 (NPR-4 which later became NPRA) by President Harding in 1923. From 1923-1926 the USGS mapped what is now NPRA and in 1930 published these results in USGS Bulletin 815, which included the first analysis of the mineral potential of the area. One of the participants of this survey was Philip Smith, Chief Alaskan Geologist of the USGS from 1925-1946. The Phillip Smith Mountains ar e named in his honor.

By 1965, industry and government had gathered over 20 years of surface geology and seismic data on the North Slope. In the 1960s industry interest focused on the area east of NPRA. Wells were drilled throughout the area between NPRA and the Arctic Wildlife Range, now the Arctic National Wildlife Refuge (ANWR).

By 1967 there were 10 plugged and abandoned exploration holes in the central arctic area. What some considered a last exploration a ttempt, and only ARCO's second well, Arco Prudhoe Bay State # 1 became the Prudhoe Bay discovery. Announced in February 1968, it came when all other North Slope oil and gas exploration activity had ceased. Today, thanks to the early explorers, the Prudhoe Bay field and the surrounding fields, account for almost 25 percent of our nation's oil production.

Gil Mull / ADGGS

Figure 90. Coastal plain area of the Arctic National Wildlife Refuge.

The National Petroleum Reserve—Alaska

NPRA , an area the size of Indiana, has not been thoroughly explored for oil and gas. Drilling and seismic data indicate that there probably is not a Prudhoe Bay size accumulation within the reserve, although smaller deposits are known to exist. The Umiat oil field may contain up to 100 million barrels of oil, but due to its isolation, it thus far has been uneconomic to develop. Most wells drilled by the U.S. Navy and the U.S. Geological Survey (mainly in the north coastal region) showed evidence of hydrocarbons. The southern part of NPRA has only been lightly explored. This area is currently virtually untouched by drilling designed to reach the type of deposits which might exist in the foothills and mountain front areas, similar to those producing oil in the Rocky Mountain region of the Lower 48. Thus, even though a Prudhoe Bay size and type of field probably does not exist , there are some 36,000 square miles which are not adequately tested for any kind of a hydrocarbon play, which may contain smaller size accumulations. Further, NPRA is known to contain large coal reserves, perhaps up to 40% of the nation's remaining reserves. Geologic samples show that southern NPRA may contain high concentrations of lead and zinc similar to the world class lead/zinc deposit at the Red Dog Mine. However, NPRA is closed to exploration for all minerals except oil and gas.

The ANWR 1002 area shares the same high potential for oil and gas with the NPRA, Prudhoe Bay, the offshore and the Mackenzie Delta Canada locales (based on the geological setting, large mapped structures and souce rock geochemistry). Surface indications of high oil and gas potential prompted Congress to set aside the 1.5 million acre 1002 area for further study (the remaining 17.5 million acres of ANWR are closed to mineral entry, including 8 million acres designated as wilderness).

Based on BLM/USGS estimates, the Department of Interior believes that the 1002 area may contain from 400 million to over 9 billion barrels of oil. ANWR also shares the same high risk, owing to uncertainties in timing of oil generation, migration, preservation, and harsh operating environment. Other considerations include high development costs and the instability associated with international supply and demand.

USDI BLM photo / Jack Petersen

Glossary and Bibliography

ablation The combined processes by which all snow or ice is lost from a glacier.

ablation moraine An irregular superglacial veneer of rock fragments formed as individual medial moraines merge, losing their identity within the glacier.

active layer The section of ground above permafrost which thaws in the summer and freezes again in winter.

allochthon A mass of rock that has been moved from its place of origin by tectonic processes, as in a thrust sheet.

alluvial fan A low, fan-shaped, relatively flat to gently sloping mass of loose rock material.

alluvium Silt, sand, and gravel deposited as a sediment during comparatively recent geologic time by a stream or other body of running water.

amphibolite A metamorphic rock consisting mainly of amphibole and plagioclase with little or no quartz.

anticline A fold, generally convex upward, whose core contains the stratigraphically older rocks.

antiform A fold whose limbs close upward in strata for which the stratigraphic sequence is not known.

arête A narrow, saw-toothed ridge created by headward erosion of cirque walls on either side of the ridge.

argillite A compact rock derived either from siltstone, claystone, or shale, that has undergone a somewhat higher degree of hardening by heat and pressure.

axis The line which, moved parallel to itself, generates the form of the fold.

basalt A dark-colored, fine-grained volcanic rock, relatively low in silica and composed predominantly of the minerals plagioclase and pyroxene.

bedding plane a planar surface that visibly separates successive layers of stratified rock.

bench gravels Gravel beds on the side of a valley above the present stream level. Represents part of a stream bed when it was at a higher level.

bentonitic A soft, plastic, porous, light-colored rock composed of clay minerals.

brachiopoda A phylum of marine, shelled animals with two unequal shells or valves.

braided stream A stream which flows in a complex of diverging and converging channels separated by sand, gravel, or mud bars. Generally thought to be caused by an inability of part of a stream to remove all of sediment load brought to it.

bryozoan Any invertebrate of the phylum Bryozoa characterized by colonial growth and calcareous skeleton. Similar to the modern sea fan.

calcareous Containing calcium carbonate (i.e., calcite).

calcite A common mineral, calcium carbonate (CaC03); the principal constituent of limestone.

carbonate rock A rock consisting chiefly of limestone or dolomite; a sedimentary rock composed of more than 50% by weight of carbonate minerals.

chalcopyrite A copper pyrite mineral (CuFeS2). An important ore of copper.

chemical weathering The process of weathering by which chemical reactions transform rocks and minerals into new chemical combinations.

chert A hard, dense, extremely fine grained sedimentary rock consisting dominantly of silica.

cirque A half-bowl or half-amphitheater carved into a mountainside by glacial erosion and frost action; the site of origin for a mountain glacier.

cirque glacier A small glacier occupying a cirque or a rounded niche formed by the ice on a mountainside.

clastic Consisting of fragments of rocks or of organic structures that have been moved individually from their places of origin.

colluvium Any loose, heterogeneous, and incoherent mass of soil material or rock fragment deposited by rainwash or slow continuous downslope movement. Usually collects at the base of gentle slopes or hillsides.

conglomerate A coarse-grained, clastic sedimentary rock composed of pebbles, cobbles, or boulders in a fine-grained matrix of sand, silt or other common cementing materials.

continental crust That type of the earth's crust that underlies the continents and continental shelves.

copper ore Ore is a naturally occurring material (minerals, in general) from which a valuable element can be extracted — in this case, copper.

coquinoid limestone A limestone consisting of coarse, unsorted, and often broken shelly materials.

cordillera A comprehensive term for an extensive series or broad assemblage of more or less parallel ranges, systems, and chains of mountains.

craton A part of the earth's crust which has had long-term structural stability and which has been little deformed for a prolonged period.

Cretaceous The final period of the Mesozoic era (after the Jurassic and before the Tertiary period of the Cenozoic era. Thought to have covered the span of time between 135 and 65 million years ago.

crevasse A deep, nearly vertical split, fissure, crack, or rift in a glacier or other mass of land ice. Caused by stresses resulting from differential movement over an uneven surface.

crossbed a single bed, inclined at an angle to the main planes of stratification.

cryoplanation Reduction of land surfaces by processes associated with frost action.

debris cone A fan shaped deposit of soil, sand, gravel, and boulders built up on a steep slope at a point where a mountain stream meets a valley.

debris flow A moving mass of rock fragments, soil, and mud, more than half of the particles being larger than sand size.

Devonian period In the ordinarily accepted classification, the fourth in order of age of the periods comprising the Paleozoic era, following the Silurian and succeeded by the Mississippian. Also the system of strata deposited at that time. Sometimes called the Age of Fishes. (345-395 million years age.)

dike A slablike, usually steeply inclined, igneous intrusive rock, which cuts across pre-existing strata.

diorite A plutonic rock (intrusive) composed essentially of sodic plagioclase (usually andesine) and hornblende, biotite, or pyroxene. Small amounts of quartz and orthoclase may be present.

dolomite A carbonate sedimentary rock consisting 50% or more of the mineral dolomite, a calcium magnesium carbonate.

drift (glacial) General term applied to material transported and deposited by a glacier.

drift (mining) A horizontal underground opening driven along a vein.

end moraine A ridge-like accumulation produced at the margin of an actively flowing glacier at any given time.

eolian Applied to deposits formed by wind deposition.

erratic (1) Large ice-borne boulders scattered so generally over the higher and middle latitudes of the Northern Hemisphere. (2) A transported rock fragment different from the bedrock on which it lies, either free or as part of a sediment. The term is generally applied to fragments transported by glacier ice or by floating ice.

extrusive Igneous volcanic rocks which have formed by cooling of magma at the earth's surface.

fault The breakages induced in once-continuous strata or structures by deformational stresses. Displacement of originally adjacent rocks occurs along such fractures or zones of fracturing.

fault gouge Soft, uncemented clay-like material found along some faults. It is formed by the crushing and grinding of rock material as the fault developed.

feldspar A group of abundant (forms 60% of the earth's crust) rock-forming minerals.

fluvial Of or pertaining to a river or rivers.

flysch A marine sedimentary environment characterized by a thick sequence of poorly fossiliferous, thin-bedded, graded deposits composed chiefly of marls, sandy and calcareous shales and muds, interbedded with conglomerates, coarse sandstones and graywackes.

fold The bending or crumpling induced in a once flat-lying structure or stratum by compressive stress.

frost action A mechanical weathering process caused by alternate or repeated cycles of freezing and thawing of water in pores and cracks.

frost wedging/splitting/shattering/riven The disintegration or splitting of jointed rock by water freezing in the cracks.

gabbro A dark-colored, coarse-grained intrusive igneous rock composed predominantly of plagioclase and pyroxene. The plutonic equivalent of basalt.

geanticline An anticlinal structure of regional extent that develops in geosynclinal (a basinlike structure) sediments due to lateral compression.

glacial lake (A) A lake that derives much or all of its water from the melting of glacial ice, as one fed by meltwater, or one lying entirely on glacial ice and due to differential melting. (B) A lake occupying a basin produced by glacial deposition as one held in by a morainal dam.

glacial outwash Stratified sand and gravel removed from a glacier by meltwater streams and deposited in front of or beyond the terminal moraine or the margin of an active glacier.

glacier A mass of ice, formed by the recrystallization of snow, that flows forward, or has flowed sometime in the past, under the influence of gravity.

glaciofluvial Pertaining to streams flowing from glaciers or to the deposits made by such streams.

gneiss A coarse-grained, high-grade metamorphic rock (see metamorphism), in which there is a rough parallelism of platy minerals combined with a prominent compositional banding of different minerals.

granite A light-colored plutonic rock consisting essentially of alkali feldspar and quartz, and dark colored minerals—usually biotite.

graywacke A dark, very hard, coarse-grained sandstone consisting of poorly sorted, angular grains of quartz and feldspar with a variety of small, dark mineral fragments embedded in a clayey matrix.

ground ice All ice of whatever origin or age, found below the ground, especially a lens, sheet, wedge, seam, or irregular mass of clear nonglacial ice enclosed in perennially or seasonally frozen ground.

ground moraine A veneer of material deposited by a glacier on the ground surface over which the glacier has moved. Forms gently rolling surface.

grus The fragmental products of in-situ granular disintegration of granitic rocks.

Holocene An epoch of the Quaternary period, from the end of the Pleistocene (about 8,000 years ago) to the present time.

hummocky ground moraine An area of knob-and-kettle topography; may have been produced by either moving or stagnant ice.

hydrocarbon Any organic compound, gaseous, liquid, or solid, consisting solely of carbon and hydrogen.

ice-contact deposits Glacial or glaciofluvial material deposited in contact with melting glacier ice, e.g., esker, kame, kettles, etc.

ice field an extensive mass of land ice covering a mountain region, consisting of many interconnected valley and other types of glaciers.

ice wedge Wedge-shaped foliated ground ice produced in permafrost, occurring as a vertical or inclined sheet, dike, or vein tapering downward.

igneous Formed by solidification from a molten or partially molten state. Rocks formed in this manner are often divided for convenience into plutonic (intrusive) and volcanic (extrusive) rocks, but there is no clear line between the two. The other two rock types are sedimentary and metamorphic (see metamorphism).

Illinoian age Pertaining to the classical third stage of the Pleistocene Epoch in North America.

intrusive/intruded Having, while fluid, penetrated into or between other rocks, but solidifying before reaching the surface.

island arc A chain of islands, e.g. the Aleutians, expressed as a curved belt rising from the deep sea floor and sometimes near the continents.

joint A surface of fracture or parting in a rock, without displacement.

Jurassic The middle one of the three periods comprising the Mesozoic era. Also the system of strata deposited during that period (193-136 million years ago).

kame A conical hill or short irregular ridge of gravel or sand deposited by a subglacial stream as a fan or delta. These features occur when a rapidly moving sediment-laden stream loses its velocity and in spreading out, deposits its load where it passes through a gap in a terminal or recessional moraine or from under an ice sheet.

kettle lake A depression in drift, made by the wasting away of a detached mass of glacier ice that had been either wholly or partly buried in the drift. A kettle which fills with water is a kettle lake.

lacustrine deposits Deposits produced by or formed in a lake.

lateral moraine (1) A low ridge-like accumulation of glacial drift deposited along the sides of a retreating glacier. (2) An elongate body of drift, commonly thin, lying on the surface of a glacier in a valley, at or near the lateral margin of the glacier. (3) An end moraine built along the lateral margin of a glacier lobe occupying a valley.

limestone A sedimentary rock made up predominantly of calcium carbonate.

lode A tabular deposit of valuable mineral between definite boundaries. A lode may consist of several veins spaced closely enough so that all of them, together with the intervening rock, can be mined as a unit.

loess Deposit of wind-blown silt or very fine sand.

ma (mega-annum) An abbreviation used in geochronologic dating (e.g., 63 Ma or 63 million years before the present).

mafic Igneous rocks (or pertaining to igneaous rocks) composed dominantly of the magnesium/iron rock-forming minerals. In general, synonymous with dark colored minerals.

marble A metamorphic rock consisting predominantly of fine-to coarse-grained recrystallized calcite and/or dolomite.

mass movement Unit movement of a portion of the land surface as in creep, landslide, or slip.

massif a massive topographic and structural feature, especially in an orogenic belt, commonly formed of rocks more rigid than those of its surroundings.

matrix Finer-grained material enclosing or filling the space between the larger grains or particles of a sediment/ sedimentary rock.

medial moraine An elongate body of drift formed by the joining of adjacent lateral moraines below the juncture of two valley glaciers.

metamorphism A process by which rocks are altered in composition, texture, or internal structure by changes in heat, pressure, and the introduction of new chemical substances.

Mississippian A period of the Paleozoic era (after the Devonian and before the Pennsylvanian) thought to have covered the span of time between 345 and 320 million years ago.

monzonite A plutonic rock containing approximately equal amounts of alkali feldspar and plagioclase, little or no quartz, and commonly augite as the main mafic mineral.

moraine A mound, ridge, or other distinct accumulation of unsorted, unstratified glacial drift, predominantly till, deposited chiefly by direct action of glacier ice.

net A form of horizontal patterned ground whose form is intermediate between a circle and a polygon.

nêvê A mass of hardened snow at the head or source of a glacier.

nunatak An isolated hill or peak which projects through the surface of a glacier. A hill or peak which was formerly surrounded but not overridden by glacial ice. An Eskimo word meaning "lonely peaks."

oceanic crust That type of the earth's crust that underlies the ocean basins. About 3.1 miles thick.

olivine An olive-green to brown rock-forming mineral of low-silica igneous rocks normally formed deep within the earth's crust.

outwash plain A broad, outspread, flat or gently sloping, alluvial sheet of material (chiefly sand and gravel) deposited by meltwater streams flowing in front of or beyond the terminal moraine of a glacier.

oxbow A closely looping stream or meander resembling the u-shaped frame embracing an ox's neck, having extreme curvature such that only a neck of land is left between two parts of the stream.

paleomagnetism Natural remnant magnetization in rocks and used to determine the intensity and direction of the earth's magnetic field in the geologic past.

patterned ground A group term for the more or less symmetrical forms such as circles, polygons, nets, steps, and stripes that are characteristic of, but not necessarily confined to, surficial material subject to intensive frost action.

pay streak The horizon or depth at which enriched mineral-bearing ground or bedrock is found.

periglacial Refers to areas, conditions, processes, and deposits adjacent to the margin of a glacier.

permafrost Permanently frozen subsoil.

phenocryst A large, conspicuous crystal set in a finer-grained groundmass of igneous rock.

phyllite A metamorphic rock of intermediate grade between slate and schist. The mica crystals impart a silky sheen to the surface of cleavage.
Pingo A large conical mound of soil-covered ice, raised by hydrostatic pressure of water within or below the permafrost layer.

pingo A large conical mound of soil-covered ice, raised by hydrostatic pressure of water within or below the permafrost layer.

placer A surficial mineral deposit formed by mechanical concentration of mineral particles from weathered debris. The mechanical agent can be alluvial, eolian, lacustrine, or glacial and the mineral is usually a heavy metal such as gold.

plate tectonics An earth model characterized by a small number of large broad plates (composed of areas of both continental and oceanic crust and mantle), each of which "floats" on some viscous underlayer. The plates float independently of the others and grind against each other. The continents form part of the plates and move with them. Crustal material is formed and expelled at mid-ocean ridges and destroyed in the ocean trenches at the continental margins.

pluton A body of igneous rock that has formed beneath the surface of the earth by consolidation from magma.

point bar A low, arcuate ridge of sand and gravel developed on the inside of a growing meander by the slow addition of individual accretions accompanying migration of the channel toward the outer bank.

polygon A form of horizontal patterned ground having 4 to 6 sides and formed through intensive frost action.

porphyritic Said of the texture of an igneous rock in which larger crystals (phenocrysts) are set in a fine-grained groundmass.

proglacial lake A lake formed just beyond the frontal margin of an advancing or retreating glacier, generally in direct contact with the ice.

Proterozoic The younger of two Precambrian systems or eras.

pyramidal peak (Horn) A high, rocky, sharp-pointed, steep-sided, pyramidal mountain peak with prominent faces and ridges bounded by the intersecting walls of three or more cirques.

pyrite A common pale-bronze or brassy yellow mineral composed of iron and sulfur.

quartz A mineral containing crystalline silica.

quartzite A metamorphic rock consisting mainly of quartz and formed by recrystallization of sandstone or chert by high pressures and temperatures, or both.

radiolaria Marine organisms (Protozoans) that possess complex internal siliceous skeletons.

recessional moraine An end moraine built during a temporary but significant halt or pause in the final retreat of a glacier.

recumbent fold An overturned fold, the axial surface of which is horizontal or nearly so.

reservoir rock Any porous and permeable rock that yields oil or gas.

rhyolite A fine-grained igneous rock having the same composition as granite.

rift A long, narrow continental trough bounded by normal faults. It marks the zone along which the entire thickness of the lithosphere has ruptured under extension.

rock glacier A glacier-like tongue of angular rock waste, cemented by interstitial ice. Usually heading in cirques or other steep-walled amphitheaters.

rocker box A cradle-like concentrating device used for sampling or mining placer deposits. Usually used when water is scarce or when placer gold is very fine.

rubble sheet Slope or valley bottom erosional landforms brought about by mass wasting in a periglacial environment.

sandstone A cemented sediment composed predominantly of quartz grains.

schist A medium or coarse-grained metamorphic rock with subparallel orientation of mica minerals which dominate its composition.

sedimentary rock A rock resulting from the consolidation of loose sediment that has accumulated in layers. Formed from fragments of older rock transported from its source and deposited in water or from air or ice. May also be formed by precipitation from solution or from the remains or sections of plants and animals.

shale A fine-grained sedimentary rock, composed of clay and silt-sized particles, which tends to break along planes parallel to bedding.

silicified carbonates The introduction of, or replacement by, silica, in limestone or dolomite (carbonates).

sill A tabular igneous intrusion that parallels the planar structure of the surrounding rock.

siltstone A compact rock having the texture and composition of shale, but lacking the fine laminations or fissility of shale.

siltstone A compact rock having the texture and composition of shale, but lacking the fine laminations or fissility of shale.

slate A slightly metamorphosed shale which tends to break or cleave along planes at an angle to the original sedimentary layering.

slickensides Polished and striated (scratched) surface that results from friction along a fault plane.

sluice box A long inclined trough formed by one or more boxes with riffles or other obstructions in the bottom for catching placer concentrates.

solifluction The process in which masses of waste, saturated with water, slowly flow from higher to lower ground. Also applied to similar subaqueous flowage. Other terms: flowing slopes, mud glaciers, mud streams.

solifluction lobe Tongue-like mass of solifluction debris commonly with steep front and relatively gentle upper surface.

source rock Sedimentary rock in which organic material under pressure, heat and time was transformed to liquid or gaseous hydrocarbons.

stagnant ice Detached blocks of ice left behind by a retreating glacier, usually buried in moraine and melting very slowly without the production of large quantities of water.

stone net A form of patterned ground whose mesh is dominantly polygonal and has a sorted appearance commonly due to a border of stones surrounding finer material.

stratified Formed, arranged, or laid down in layers or strata.

stream terrace Stream-cut rock terrace with thick cover of slope wash.

strike slip fault A fault on which the horizontal displacement is parallel to the fault's strike.

syenite A plutonic rock containing alkali feldspar, a small amount of plagioclase, one or more mafic minerals, and quartz, if present, only as an accessory.

syncline A fold of which the core contains the stratigraphically younger rocks; it is generally concave upward.

talus Rock fragments of any size or shape derived from and lying at the base of a cliff or very steep, rock slope.

tectonics Study of the broader structural features of the earth and their causes.

terminal moraine A moraine formed across the course of a glacier at its farthest advance, at or near a relatively stationary edge, or at places marking the termination of important glacial advances.

terminus The lower margin or extremity of a glacier.

terrace Any long, narrow, relatively level or gently inclined surface bound on one end by a steeper descending slope and along the other by a steeper ascending slope.

terrane A faulted-bounded geologic entity characterized by a distinctive geologic history that differs markedly from the geologic history of adjacent areas.

Tertiary The earlier of the two geologic periods comprising the Cenozoic era. Also, the system of strata deposited during that period (1.6 - 66.4 million years ago).

thermo-erosional niching Undercutting of ice rich soils produced by bank erosion in regions containing permafrost.

thermokarst lake/pond/pit Caved-in lakes, bogs, caverns, pits and other small depressions formed in permafrost regions by the melting of ground ice.

thrust fault A fault in which the fault plane is inclined at a particularly shallow angle (less than 45°).

till Unsorted and non-layered glacial deposits laid down directly by the ice.

Triassic The earliest of the three geologic periods comprising the Mesozoic era. Also the system of strata deposited during that period (208-245 million years ago).

troy ounce A unit of weight equal to 31.1 grams or 1.1 ounces avoirdupois (there are 12 troy ounces in one troy pound and 16 avoirdupois ounces in one avoirdupois pound). Used when expressing the weight of precious metals.

tuff A rock formed of compacted volcanic fragments, generally smaller than 0.2 inches in diameter.

turbidite A sediment or rock deposited from a turbidity current; a bottom-flowing current laden with suspended sediment, moving swiftly down a subaqueaous slope and spreading horizontally on the floor of the body of water.

ultramafic Said of igneous rocks composed chiefly of mafic minerals.

unconformity A gap in the geologic record where a rock unit is overlain by another that is not next in stratigraphic succession.

U-shaped valley A valley having the form of a broad letter U with steep parallel walls and a broad, nearly flat floor. Specifically, a valley carved by glacial erosion.

Utility Corridor The Dalton Highway lies within the Utility Corridor, which is 304 miles long and from 6 to 24 miles wide. The corridor consists of 2.7 million acres of public land surrounding the pipeline.

vein A mineral filling of a fracture in a host rock; often associated with mineral deposits.

valley glacier A glacier flowing between the walls of a mountain valley in all or part of its length.

weathering A destructive process by which earth and rock materials, upon exposure to atmospheric agents at or near the earth's surface, change in color, texture, composition, firmness, or form with little or no transport of the loosened or altered material.

Wisconsinan age Fourth Pleistocene glaciation (approximately 12,000 years ago).

References Cited

Beikman, H. M., 1980, Geologic map of Alaska: U.S. Geological Survey, scale 1:2,500,000, 2 sheets.

Brosge', W. P., and Reiser, H. N., 1964, Geologic map and section of the Chandalar quadrangle, Alaska: U.S. Geologic Survey Miscellaneous Geologic Investigations Map I-375, scale 1:250,000.

Brosge', W. P., and Reiser, H. N., 1971, Preliminary bedrock geologic map of Wiseman and eastern Survey Pass Quadrangles, Alaska: U.S. Geological Survey Open-file Report 71-56, scale 1:250,000.

Brosge', W. P., Reiser, H. N., Dutro, J. T., Jr., and Detterman, R. L., 1979, Bedrock geologic map of the Philip Smith Mountains quadrangle, Alaska: U.S. Geological Survey Miscellaneous Field Studies Map MF-879-B, scale 1:250,000.

Brosge', W. P., Reiser, H. N., and Yeend, W., 1973, Reconnaissance geologic map of the Beaver quadrangle, Alaska: U.S. Geological Survey Miscellaneous Field Studies Map MF-525, scale 1:250,000.

Brown, J., and Berg, R. L., 1980, Environmental engineering and ecological baseline investigations along the Yukon River-Prudhoe Bay Haul Road: U.S. Army Cold Regions Research and Engineering Laboratory Rept. 80-19, 187 p.

Brown, J., and Kreig, R. A., eds, 1983, Guidebook to permafrost and related features along the Elliot and Dalton Highways, Fox to Prudhoe Bay, Alaska: Alaska Division of Geological and Geophysical Surveys Guidebook 4, 230 p.

Chapman, R. M., Weber, F. R., and Taber, B., 1971, Preliminary geologic map of the Livengood quadrangle, Alaska: U.S. Geological Survey Open-file Report 71-66, scale 1:250,000, 2 sheets.

Coulter, H. W., Hopkins, D. M., Karlstrom, T. N. V., Pewe, T. L., Wahrhaftig, Clyde, and Williams, J. R., 1965, Map showing extent of glaciations in Alaska, compiled by the Alaska Glacial Map Commitee: U.S. Geological Survey, Miscellaneous Investigations Map I-415, scale 1:2,500,000.

Dillon, J. T., Brosge', W. P., and Dutro, J. T., 1986, Generalized geologic map of the Wiseman quadrangle, Alaska: U.S. Geological Survey Open-file Report 86-219, scale 1:250,000, 1 sheet.

Hamilton, T. D., 1978a, Surficial geologic map of the Philip Smith Mountains quadrangle, Alaska: U.S. Geological Survey Miscellaneous Field Studies Map MF-879-A, scale 1:250,000.

Hamilton, T. D., 1978b, Surficial geologic map of the Chandalar quadrangle, Alaska: U.S. Geological Survey Miscellaneous Field Studies Map MF-878-A, scale 1:250,000.

Hamilton, T. D, 1979b, Surficial geologic map of the Wiseman quadrangle, Alaska: U.S. Geological Survey Miscellaneous Field Studies Map MF-1122, scale 1:250,000.

Hamilton, T. D., 1982, A late Pleistocene glacial chronology for the southern Brooks Range: Stratigraphic record and regional significance: Geological Society of America Bulletin, v. 93, p. 700-716.

Hamilton, T. D., 1983, Glacial geology of the Brooks Range, *in* Brown, J., and Kreig, R. A., eds., Guidebook to permafrost and related features along the Elliot and Dalton highways, Fox to Prudhoe Bay, Alaska: Alaska Division of Geological and Geophysical Surveys Guidebook 4, p. 22-25.

Hamilton, T. D., and Hopkins, D. M., 1982, Correlation of northern Alaskan glacial deposits-a provisional stratigraphic framework, *in* Coonrad, W. L., ed., The United States Geological Survey: Accomplishments during 1980: U.S. Geological Survey Circular 844, p. 15-18.

Holmes, G. W., Hopkins, D. M., and Foster, H. L., 1968, Pingos in central Alaska: U.S. Geological Survey Bulletin 1241-H, p. H1-H40.

Huffman, A. C., Jr., ed., 1985, Geology of the Nanushuk Group and related rocks, North Slope, Alaska: U.S. Geological Survey Bulletin 1614, 129 p.

Jones, D. L., Silberling, N. J., Coney, P., and Plafker, G., 1984b, Lithotectonic terrane map of Alaska (west of the 141st meridian), *in* Silberling, N.J., and Jones, D.L., eds, 1984, Lithotectonic terrane maps of the North American Cordillera: U.S. Geological Survey Open-file Report 84-523, 12 p., scale 1:2,500,000, 4 sheets.

Judson, S., Deffeyes, K. S., and Hargraves, R. B., 1976, Physical Geology: Prentice-Hall, Inc., Englewood Cliffs, New Jersey, 560 p.

Kreig, R. A., and Reger, R. D., 1982, Air-photo analysis and summary of landform soil properties along the route of the trans-Alaska pipeline system: Alaska Division of Geological and Geophysical Surveys Geologic Report 66, 149 p.

Kudriavtsev, V. A., 1978, Obchchee Merzlotovedeniit - Geokrylogiia Izd. Moskovskogo Universiteta, 463 p. (General Permafrost Science - Geocryology.)

Lachenbruch, A. H., 1962, Mechanics of thermal contraction cracks and ice-wedge polygons in permafrost: Geological Society of America Special Paper 60, 69 p.

Linnell, K. A., 1973, Long term effects of vegetation cover on permafrost stbility in an area of discontinuous permafrost: Proceedings of the Second International Conference on Permafrost, Yakutsk, USSR, North American Contribution, p. 688-693. Washington, D.C., National Academy of Science.

Marshall, Robert, 1933, Arctic Village: New York, Smith and Haas, 399 p.

Mull, C. G., Roeder, D. H., Tailleur, I. L., Pessel, G. H., Grantz, Arthur, and May, S. D., 1987, Geologic sections and maps across Brooks Range and Arctic slope to Beaufort Sea, Alaska: The Geological Society of America Map and Chart Series MC-28S, 1 sheet.

Patton, W. W., and Miller, T. P., 1973, Bedrock geologic map of Bettles and southern part of Wiseman quadrangles, Alaska: U.S. Geological Survey Miscellaneous Field Studies Map MF-492, scale 1:250,000.

Rickwood, F. K., 1970, The Prudhoe Bay field *in* Proceedings of the Geological Seminar on the North Slope of Alaska: American Association of Petroleum Geologists.

Silberling, N. J., and Jones, D. L., eds., 1984, Lithotectonic terrane maps of the North American Cordilleran: U.S. Geological Survey Open-file Report 84-523, 12 p., scale 1:2,500,000, 4 sheets.

Specht, R. N., Brown, A. E., Selman, C. H., and Carlisle, J. H., 1987, Geophysical case history, Prudhoe Bay field, *in* Tailleur I., and Weimer, P., eds., Alaskan North Slope geology: Alaska Geological Society, Anchorage, Alaska, p. 19-39.

Strahler, Arthur N., 1963, The Earth Sciences.

Walker, H. J., 1973, Morphology of the North Slope, in Britton, M. E., ed., Alaskan arctic tundra: Arctic Institute of North America Technical Paper No. 25, p. 49-92.

Suggested Reading

Banet, A. C., Jr., 1991, Oil and gas development on Alaska's North Slope: Past results and future prospects: U.S. Bureau of Land Management, Alaska State Office Open-file Report 34, 42 p.

Beistline, E. H., 1966, Placer mining in frozen ground, in International Conference on Permafrost, Lafayette, Ind. 1963, Proceedings: National Academy of Sciences - National Research Council Publication 1287, p. 463-467.

Brouwers, E. M., Clemens, W. A., Spicer, R. A., Ager, T. A., Carter, L. D., Sliter, W. V., 1987, Dinosaurs on the North Slope, Alaska: High latitude, latest Cretaceous environments: Science, v. 237, p. 1608-1610.

Carson, C. E., and Hussey, K. M., 1962, The oriented lakes of arctic Alaska: Journal of Geology, v. 70, no. 4, p. 417-439.

Davies, K. L., 1987, Duck-bill dinosaurs (Hadrosauridae; Ornithischia) from the North Slope of Alaska: Journal of Paleontology v. 61, p. 198-200.

Detterman, R. L., Bowsher, A. L., and Dutro, J. T. Jr., 1958, Glaciation of the Arctic Slope of the Brooks Range, northern Alaska: Arctic, vol. 11, no. 1, p. 43-61.

Dillon, J. T., 1982, Source of lode and placer gold deposits of the Chandalar and upper Koyukuk districts, Alaska: Alaska Division of Geological and Geophysical Surveys Open-file Report 158, 22 p., scale 1:250,000, 1 plate.

Embleton, C., and King, C., 1975b, Periglacial geomorphology: John Wiley and Sons, New York, 250 p.

Hamilton, T. D., Obi, C. M., 1982, Pingos in the Brooks Range, northern Alaska, U.S.A.: Arctic Alpine and Research, v. 14, no. 1, p. 13-20.

Jamison, H. C., Brockett, L. K., and McIntosh, R. A., 1980, Prudhoe Bay—a ten year perspective, in Halbounty, M. T., ed., Giant oil and gas fields of the decade 1968-1978: American Association of Petroleum Geologists Memoir 30, p. 289-314.

Marshall, Robert, 1934, Reconnaissance of the northern Koyukuk valley, Alaska: U.S. Geological Survey Bulletin 844-E, p. 247-256.

Mull, C. G., 1972, Leffingwell..Pioneer arctic explorer-scientist: Alaska Geographic, v.1, no. 1, p. 56-58.

Mull, C. G., 1982, History of Arctic Slope oil exploration; Alaska's Oil/Gas and Minerals Industry: Alaska Geographic, v. 9, no. 4, p. 188-198.

Mull, C. G., and Harris, E. E., 1989, Dalton Highway road log from Chandalar Shelf to Prudhoe Bay, *in* Mull, C.G., and Adams, K. E., eds., Dalton Highway, Yukon River to Prudhoe Bay, Alaska: Alaska Division of Geological and Geophysical Surveys Guidebook 7, p. 101-132.

Patton, W. W., Jr., Miller, T. P., Box, S. E., 1989, Dalton Highway road log from Yukon crossing to South Fork Koyukuk River, *in* Mull, C.G., and Adams, K. E., eds., Dalton Highway, Yukon River to Prudhoe Bay, Alaska: Alaska Division of Geological and Geophysical Surveys Guidebook 7, p. 59-73.

Pe'we', T. L., 1975, Quaternary geology of Alaska: U.S. Geological Survey Professional Paper 835, 145 p.

Tuck, Ralph, 1940, The origin of muck-silt deposits at Fairbanks, Alaska: Geological Society of America Bulletin, v. 51, p. 1295-1310.

Wahrhaftig, Clyde, and Cox, Allan, 1959, Rock glaciers in the Alaska Range: Geological Society of America Bulletin, v. 70, no. 4, p. 383-436.

Washburn, A. L., 1973, Periglacial processes and environments, New York, St. Martin's Press, 320 p.

White, S. E., 1976, Rock glaciers and block fields, review and new data: Quaternary Research, vol. 6, p. 77-97.

Will, Susan M., 1981, Coldfoot: An historic mining community on the Middle Fork of the Koyukuk River, Alaska: Unpublished Report, U.S. Bureau of Land Management, Fairbanks District Office, 88 p.

The BLM Minerals Program

Few people realize that the Bureau of Land Management minerals program evolved over 100 years and involved three agencies, the General Land Office, the Bureau of Mines and the U. S. Geological Survey. The roots of the program go back to the mining laws of 1866, 1870, 1872 and 1873. These laws effectively opened the west to minerals development by proclaiming the public lands open for private mineral exploration and giving property deeds to owners of valid claims. Congress passed the Minerals Leasing Act in 1920 and changed the disposal of oil, gas and coal from "location" under the mining laws to competitive leasing. These minerals became known as "leasables."

The various minerals responsibilities for administration, oversight and land classification under these acts, were divided between the General Land Office, the Bureau of Mines Leasing Division, and the U.S. Geological Survey Lands Classification Branch (later the Conservation Division of the USGS), respectively.

Today the Energy and Minerals Resources program within BLM has many of the onshore minerals responsibilities formerly found within these agencies. These include: leasing and lease management of minerals on all federal onshore lands; inspection and enforcement of all oil and gas activities on public onshore lands to ensure compliance with federal environmental laws; recording, managing, patenting, and inspecting federal mining claims; evaluating the mineral potential of BLM and; identifying how these minerals can best be managed under multiple use; and identifying and evaluating subsurface energy minerals to determine their economic value.

The Bureau of Land Management balances careful exploration and development of resources with sound environmental practices under the legislative mandate of sustained yield for future generations. In Alaska, the BLM Division of Mineral Resources also monitors the trans-Alaska pipeline right-of-way agreements, works closely with the U. S. Fish and Wildlife Service on minerals evaluations and oil and gas management in the national wildlife refuges, and evaluates the recoverable oil and gas resource potential of the Arctic National Wildlife Refuge coastal plain area.